THE OAKWOOD LIBARY OF RAILWAY HIST

The South Shields, Marsden & Whitburn Colliery Railway

by

William J. Hatcher

THE OAKWOOD PRESS

British Library Cataloguing in Publication Data
A Record for this book is available from the British Library
ISBN 0 85361 583 7

Typeset by Oakwood Graphics.
Repro by Ford Graphics, Ringwood, Hants.
Printed by Cambrian Prinetrs Ltd, Aberystwyth, Ceredigion.

On 15th September, 1949, the ex-NER 'C' class engine No. 8 crosses Lighthouse bridge with a Westoe Lane-bound service. This track was in fact the 'main line' with the other road employed as a mineral loop into the pit sidings. The chimney of Marsden village school can be seen beyond the bridge. *P.J. Lynch*

Front cover: Whitburn's Pride. The ex-NER 'C' class engine No. 8 appears to have been freshly coaled and watered and is running through the Bank sidings at Whitburn Colliery ready for another turn of duty on 1st April, 1934. An SSMWCR number plate is visible on the cabside supplemented by the lettering HCC and No. 8 painted on the tender. *H.C. Casserley*

Title Page: Whitburn Colliery station on 17th April, 1968, 15 years after closure, 21 ton steel-bodied wagons stand beneath the No. 1 winder while earlier wooden-bodied stock on the left occupy the long-redundant carriage siding. Note the platform lighting and Marsden limestone walling. *Ian S. Carr*

Published by The Oakwood Press (Usk), P.O. Box 13, Usk, Mon., NP15 1YS.
E-mail: oakwood-press@dial.pipex.com
Website: www.oakwood-press.dial.pipex.com

Contents

Robert Stephenson & Hawthorn No. 7603 was the first of the 'Stubby Hawthorns' to arrive on the Marsden Railway. With six Great North of Scotland Railway carriages in tow, No. 7603 is just leaving the original course of the SSMWCR and entering the re-aligned section through the Grotto Cutting. The fenced off Coast Road is running alongside in this 1950s photograph. *Neville Stead Collection*

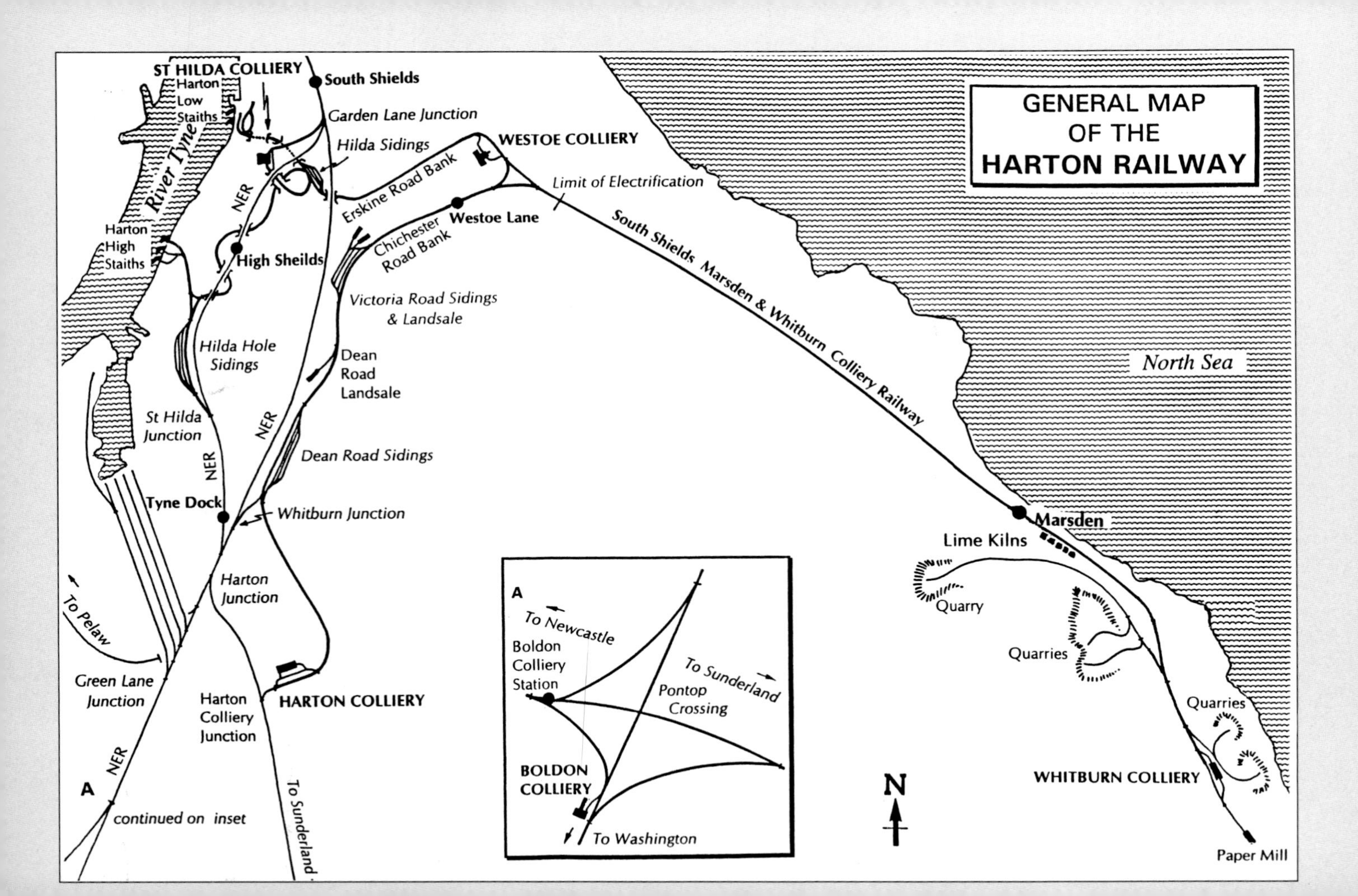
GENERAL MAP
OF THE
HARTON RAILWAY
ST HILDA COLLIERY
Harton Low Staiths
South Shields
Garden Lane Junction
Hilda Sidings
WESTOE COLLIERY
River Tyne
NER
Erskine Road Bank
Limit of Electrification
Westoe Lane
Harton High Staiths
High Sheilds
Chichester Road Bank
South Shields Marsden & Whitburn Colliery Railway
Victoria Road Sidings & Landsale
Hilda Hole Sidings
Dean Road Landsale
North Sea
St Hilda Junction
Dean Road Sidings
Tyne Dock
Whitburn Junction
Marsden
Lime Kilns
Harton Junction
Quarry
To Pelaw
Quarries
Green Lane Junction
Harton Colliery Junction
HARTON COLLIERY
A
To Newcastle
Boldon Colliery Station
To Sunderland
Pontop Crossing
BOLDON COLLIERY
To Washington
N
WHITBURN COLLIERY
To Sunderland
continued on inset
Paper Mill

Introduction

The South Shields, Marsden & Whitburn Colliery Railway (SSMWCR), this grand rather over-explanatory title was bestowed upon a section of track just 2¾ miles from end to end. The SSMWCR was originally built as a mineral branch serving a coal mine on the North-East coast, and this it did successfully for almost a century. Indeed here is where the story could have ended, for at face value this was no different from the hundreds of other branches which served pits in this industrial corner of the country.

Scratch beneath the surface however and it soon becomes apparent that there was much more to the SSMWCR than the humble coal truck. The railway had its own passenger service which (although not unique for a colliery branch line) nonetheless boasted well-equipped termini and block signalling. Passenger trains (which ran at a loss for much of their existence) were capable of taking the public back in time, leaving Westoe Colliery at one end of the branch, with its 20th century electric railway, and travelling down to Whitburn Colliery at the other end, firmly set in the 19th century with steam the staple power.

Then there was the rolling stock. This bewildering menagerie was raked in from all corners of the land, mostly via the second-hand market and spent a twilight existence on the branch line long after similar main line stock had been dragged to the scrapyards. The locomotives (and there were 40 of them) represented a full cross-section of early mineral designs from tiny Manning, Wardle industrial tank engines up to the hefty ex-North Eastern Railway (NER) 'C' class tender engines which were so indigenous to County Durham.

The most remarkable aspect of the railway, however, was the state of almost constant change it underwent throughout its history. It was built under wayleave as a mineral line, re-opened by Act of Parliament as a railway in the full sense of the word, and re-opened yet again at a later date as an official Light Railway. The original stations were demolished and rebuilt, signal boxes appeared on the railway, only to disappear again and be replaced by other boxes, while at one stage, even part of the railway itself was lifted and re-aligned. In fact so frequent were the changes that an enthusiast visiting the railway could return within a decade and find little to remind him of his previous visit.

The following text is an attempt to describe and encompass all the components which combined to make the SSMWCR such a fascinating subject.

The NER 'C' class locomotive No. 8 has just crossed Lighthouse Bridge with a 'Rattler' service comprising ex-North British Railway stock and will shortly terminate at Whitburn Colliery on 1st April, 1934. The Souter Point lighthouse is just visible to the right of the tender.
H.C. Casserley

Contrary to this deserted 1904 view, Marsden Bay was a popular holiday resort at the time, though this popularity was never really capitalised on by the Harton Coal Co. Clearly seen are the Marsden Rock, and Marsden Grotto, while it is also just possible to make out on the cliff top, (*from left to right*) Marsden village,the Lighthouse Quarries, and a rake of railway wagons.

Public Record Office, Durham/British Coal

Chapter One

The Formative Years

Although the South Shields, Marsden & Whitburn Colliery Railway is remembered for its little passenger trains, it was coal, of course, which brought about the existence of the line.

Coal had been mined in South Shields since the Roman occupation, and the land pitted over the centuries by small scale operations. Industrialisation did see the establishment of one or two larger undertakings, but it took the formation of the Harton Coal Company (HCC) in 1842 to bring about an integrated coal industry to South Shields. The brainchild of Nicholas Wood and Christopher Blackett, the HCC leased a vast tract of land from the existing land owners (the 'Dean and Chapter of Durham') and this gave the company exclusive rights to extract coal from a triangle of land formed by the north-eastern tip of County Durham, Jarrow Slake on the edge of the Tyne, and Whitburn on the coast.

The HCC immediately purchased an existing pit, in the centre of South Shields, St Hilda Colliery, and began construction of a second at Harton to the south of the town. Harton Colliery produced its first coals on 8th July, 1844 with the coal company's third pit, Boldon, opening for business on 20th July, 1869.

These three pits were capable of extracting all inland coal from within the HCC domain, and the company now looked greedily toward the rich reserves which lay offshore. Indeed, these very reserves were the reason why Wood and Blackett chose to build their empire at South Shields in the first place. Earlier borings conducted on the headland at Benthouse and Marsden back in 1779 promised a treasure trove of thick coal seams beneath the sea bed, and the HCC put together a plan in 1873 to sink two coastal pits some three miles apart.

The year 1874 saw the Harton Coal Company cleave its empire (on what was to be a temporary basis) into two separate businesses overseen by the parent company. The Harton & Hilda Coal Company managed the established pits at Harton, St Hilda, and Boldon, while the Whitburn Coal Company (WCC) was to instigate the construction of the new coastal mines.

Problems were immediately encountered during the sinking of the shaft at the new Whitburn Colliery, and these were of such magnitude that construction of the second pit at Benthouse was immediately put on hold. During the sinking, the engineers had unleashed a hidden chasm of water and quicksand forcing them to tunnel a lateral drift through the cliffside in an attempt to drain off the water. A newspaper account of the day graphically illustrates the catastrophe:

> . . . nor did the influx come from the sea alone. The land round about contributed a large quota of fresh water which met the saltwater in the sinkings and made common cause against the undertaking. The drift ran like a mountain torrent. The water, having run the course of the drift, fell over the rocks and into the sea like a veritable Niagara, and hundreds of people have travelled to witness the spectacle.

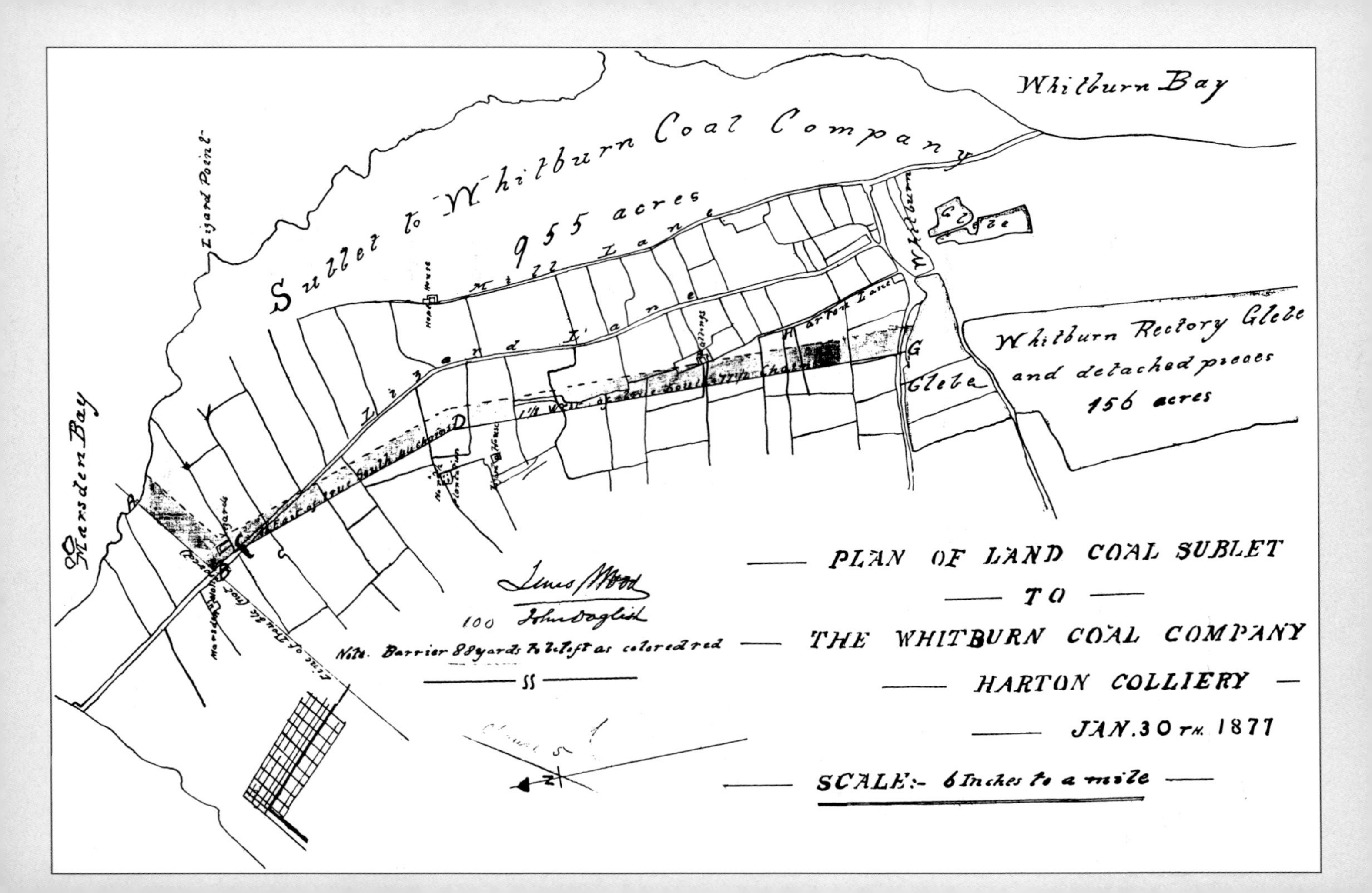

Whitburn Bay
Sublet to Whitburn Coal Company
955 acres
Lizard Point
Marsden Bay
Whitburn Rectory Glebe
and detached pieces
156 acres
Glebe
Glebe
Lizard Lane
Mill Lane
Harton Lane
Note. Barrier 88 yards to be left as colored red
PLAN OF LAND COAL SUBLET
TO
THE WHITBURN COAL COMPANY
HARTON COLLIERY
JAN. 30TH 1877
SCALE:- 6 Inches to a mile

At a cost of £15,000, the WCC called upon engineers from Belgium who were expert in this field. They employed the King-Chandron method whereby hoops of cast-iron tubing were inserted into the shaft reducing its bore from 16 ft to 12 ft diameter. Three pumps were then used to extract the water from within the lined shaft at an incredible 12,000 gallons per minute.

Flood waters aside, the WCC had other difficulties to contend with, the most significant being the transportation of coal from this desolate stretch of coastline down to the River Tyne. Harton Colliery, complete with its rail link with the North Eastern Railway lay due west, and the most obvious course would have been to construct the line in this direction. However, a long bank of high ground known as the Cleadon Hills provided a natural barrier against this route forcing the engineers to forge northwards to the village of Westoe on the edge of South Shields. From here a curve would send the railway inland in a south-westerly direction, eventually linking it up with the NER north of Tyne Dock at what would become Whitburn Colliery Junction.

The WCC was granted permission to build this 3 mile 7 furlong 1½ chain single-track railway by the Dean and Chapter of Durham and the line was thus built by wayleave rather than by Act of Parliament. In fact the coal company leased several large tracts of land totalling 465 acres at Marsden and a great change swept across the district in the 1870s.

Before the coming of the mine, Marsden consisted of a scattering of farmsteads, lonely cottages, and a lighthouse (Souter Point - built 1871) yet the end of the century would see the population swell to 3,750. One small existing industry in the area had been limestone production and this had been quarried on a local scale for building purposes, or burnt to produce lime for the farming community. Five such quarries were purchased by the Whitburn Coal Co. in 1874. Of these five, the two immediately east of the Whitburn Colliery site (the so-called Harbour Quarries) were already exhausted and were earmarked for the disposal of spoil from the pit, while a third face, the Marsden Old Quarry was a small, remote excavation of limited potential. However, the fourth and fifth excavations (which became known as the Lighthouse Quarries) were far more exciting propositions for the WCC. Lying adjacent to the new railway, and ¼ mile to the north of the pit, the stone here was found to be in great quantity and largely unworked. The coal company set up a horse-worked, 2 ft gauge track within both quarries, and built a large battery of lime kilns into the hillside which overlooked both the new railway and the sea.

All this new industry required a work force of course, and to cater for this the WCC built from scratch the new village of Marsden. Conveniently situated between the Lighthouse Quarries and Whitburn Colliery, Marsden village consisted of 134 houses of either 'two-up two-down' or 'one-up two-down' design laid out in nine streets which catered for some 700 people. There was a post office-cum-general store (with a backroom where the doctor and dentist practised), a union room (which doubled as a barber shop), a Methodist Chapel, an ambulance hall, and a recreation field. Added to this was a school for 270 children in 1882, a miner's reading room in 1896 (instigated to prevent miners 'loitering around in the streets') and St Andrew's church in 1900. The village was situated somewhat precariously on the clifftop and life there must have been bleak to say the least,

Harton Coal Company and The Whitburn Coal Company Lim:

Articles of Agreement

made the thirtieth day of December One thousand eight hundred and seventy eight Between The Whitburn Coal Company Limited who are hereinafter referred to as the Company of the first or one part and Collingwood Lindsay Wood of Keeland in the county of Perth Esquire Lindsay Wood of Southill in the county of Durham Esquire John Wood of 89 Gloucester Terrace in the county of Middlesex Esquire Nicholas Wood of 35 Albermarle Street in the same county Esquire, Christopher Edward Blackett at present residing at Exeter Grange in the county of Rutland a Colonel in Her Majesty's Army Charles William Anderson of Kirkhamerton Hall in the county of York Esquire and Hilton Philipson of the Borough and county of Newcastle upon-Tyne Esquire (carrying on business together under the style or firm of the Harton Coal Company) of the second or other part Whereas the parties hereto are as lessees under the Ecclesiastical Commissioners for England entitled to wayleave and liberty of passage and other privileges over and along the strip or parcel of ground situate at or near Westoe in the county of Durham and colored red on the plan annexed to this agreement and signed by John Daglish one of the Directors of the Company and the said Lindsay Wood for a term of years granted or agreed to be granted to them by the said Commissioners at and under the rent or rents and subject to the covenants conditions and agreements to be paid performed and observed by them in respect thereof and they are as such lessees entitled to exercise the right of constructing a railway or wagonway with the byeways sideways branches bridges tunnels mounds batteries depôts and other works incidental or appertaining thereto over and upon the said piece of ground And Whereas the said parties hereto of the second are seized of the piece or parcel of ground situate at Westoe aforesaid and shewn on the said plan and thereon coloured green for an estate of inheritance in fee simple in possession

The leading page of the articles of agreement for running the Marsden Railway as drawn up by the Harton Coal Co. and the Whitburn Coal Co. on 13th December, 1878.

particularly so when the village was being battered by one of the notoriously savage North Sea gales which still frequent the coast to this day. Little wonder that many Whitburn miners preferred the sheltered civilisation of South Shields.

With construction of the railway now well under way, 1878 saw the signing of two agreements vital to its completion.

Although not signed until November 1878, the first agreement came before the South Shields Corporation in June of that year. Officially known as the Whitburn Agreement, this was basically a hammering out of differences with much give and take required by both Council and coal company. The impending Benthouse Colliery plans were embroiled with those for the new railway and this complicated matters further. However, an agreement was finally reached, with the WCC given permission to construct the final section of its railway through South Shields in return for it rebuilding parts of re-routed streets (namely Ravensbourne Terrace, Mowbray Road and Erskine Road). The Corporation also insisted that all railway bridges in the town area be constructed to requirements which stipulated that they must be both watertight and of ornamental appearance.

The second agreement involved the WCC and its parent company the HCC, and this was officially titled the Harton Coal Company and Whitburn Coal Company Union Railway Agreement. Drawn up on 13th December, 1878 this consisted of 21 articles which effectively bound the two companies to certain duties. The articles of agreement were drawn up in a 30 page document too convoluted to be reproduced here, however some interesting detail can be gleaned from the text, and this detail is summarised below:

The WCC could build the railway on HCC land but at the sole expense of the WCC.
The railway had to be built within 12 months of the above date.
Coal depots were to have at least 8 cells and 2 sidings.
Construction costs were to be borne between both companies.
Neither company was to sell coal at a lower price than the other company.
All rents paid to the Ecclesiastical Commissioners by both companies were according to tonnages carried.
Neither company was to allow an outside company to use the railway without the other's permission.

This so called Union Railway Agreement was then signed by both Boards of Directors these being:

For the Harton Coal Co.
Nicholas Wood (Chairman), Collingwood Wood, Lindsay Wood, John Wood, Christopher Blackett, Charles Andover, and Hilton Phillipson.

For the Whitburn Coal Co.
Hilton Phillipson (Chairman), George Butler, John Butler, George Campbell, Eric Cooper, John Dalgleish, Newton Clark.

It is interesting to note that the Chairman of the Whitburn Coal Co. was also on the Harton Coal Co. Board of Directors revealing just how closely interwoven the two companies were.

Whitburn Colliery and its new railway were officially opened on 1st May, 1879. However, no mention was made of this by the press of the day regarding an official opening ceremony. In fact no official name was bestowed on the railway at the time, although it was referred to locally as both 'the Marsden Railway' and 'the Whitburn Colliery Branch' (indeed, the pit itself was referred to as either 'Whitburn' or 'Marsden'). This rather inauspicious beginning was due to the fact that Whitburn pit was not ready to start full scale operations. In fact, the first shaft at Whitburn was only 420 ft deep by November and even as late as 1881 the pit only produced 9,222 tons of coal for the entire year.

It was limestone therefore, that was carried initially on the railway. This was in the form of dimension stone for building or masonry work, screened limestone for the steel industry, and lime for the farming community, all hauled down to Whitburn Colliery Junction for exchange with the NER by the only two locomotives then extant on the railway (a pair of 0-6-0 saddle tanks of Manning, Wardle and Black, Hawthorn extraction).

The workforce involved in both the construction and working of the pit, quarries, and village was considerable and had to travel to and from South Shields on a daily basis. Because there was no direct road between South Shields and Marsden at the time, presumably this workforce travelled by company train. However, there is no hard evidence to back this up, the earliest reference to passengers not appearing until 1885, six years after the railway opened. On 13th November of this year, the Railway Department of the Board of Trade received the following letter from the Whitburn Coal Company solicitors Phillipson, Cooper and Goodger:

Sir,

Our clients the Whitburn Coal Company are the owners of a large Colliery and works situated some 3 or 4 miles from the town of South Shields and they are also the owners of considerable landed property, and partly on their property and partly on another property, by virtue of a wayleave bill, they have constructed a railway which passes through the town of South Shields to join the public line of railway which leads to the shipping places on the river Tyne.

A large number of their workmen reside in, or near to, the town of South Shields and they, somewhat naturally, object to travel so great a distance to their work.

In addition to this their friends and other persons who visit the Colliery find it a very great inconvenience to travel on the wagonway inasmuch as the road is circuitous.

It appears to us however that charges may possibly come under the description of 'passengers' and that their conveyance may render it necessary for the Company to have their line inspected by the Board of Trade and to have their Certificate under the Act: 506 VICTORIA cap 55 s 21 [*sic*].

We are therefore instructed by the Whitburn Coal Company to bring the matter before you to ask you to be good enough to inform us what, if any preliminaries are requisite before they can be placed in a position to convey persons whether workmen or not along their line to and from their property in the neighbourhood of which we may mention, there are already a number of houses.

Yours obediently

Phillipson Cooper Goodger

This somewhat rambling letter does not quite say whether passengers (in the form of workmen) were already being conveyed on the railway or not, and just as curious was the fact that the matter was still not resolved some 19 months later! The following letter was written on 20th June, 1887 by Major General C.S. Hutchinson of the Board of Trade to his Head Office in Whitehall and is reproduced in full as it contains a description of the line in its original form.

Sir,

I have the honour to report, for the information of the Board of Trade, that in compliance with the instructions contained in your minute of the 11th instant, I have inspected a portion of the Whitburn Coal Company's Railway.

The portion of this line was presented for inspection with a view to its being used for passenger traffic extends from South Shields to Marsden, a length of two and three quarter miles.

The line is single, with sidings at the termini, the gauge 4 ft 8½ inches, the space between the lines where there is more than one is 6 ft and the width at formation is 13 ft. Land has been purchased for a second line of rails.

The steepest gradient has an inclination of 1 in 185, and the sharpest curve a radius of 17 chains.

The permanent way consists of double headed steel rails in 24 ft and 30 ft steel lengths weighing 82 lbs per yard, fished at the joints, and supported in cast iron chairs weighing 40 lbs secured by spikes and keyed on the outside. The sleepers are creosoted 9 ft long by 10 x 5 inches placed one yard from centre to centre.

The ballast is of boiler ashes, 6 inches deep below the under surface of the sleepers.

There are 3 bridges under the line constructed with stone abutments and wrought iron girders widest span 40 ft.

Over the line is a footbridge of timber, span 28 ft.

These bridges (with the exception of that at 2 miles 3 chains where a bottom plate is required to the lower flange of the girder) seem to be substantially constructed and to be standing well. The girders of the wide bridges (with the above exception) have sufficient theoretical strength and gave moderate deflections when tested.

The fencing is of post and rail and of iron handles.

There are at present no station buildings or interlocking arrangements.

I observe the following requirements:

1. The permanent way requires improvement. More ballast is necessary and the sleepers should be equally spaced and set square to the rails.
2. Trusses and ties are required at the wide bridges, and a bottom plate to the girder of the bridge at 2 miles 3 chains.
3. Proper signalling arrangements, shelter, booking offices, and conveniences are required at the stations.
4. Gradient boards and mileposts should be supplied.
5. The formation is narrow on some of the banks and should be increased.
6. Some additional fencing is required.
7. Some gateposts are too close to the line and must be set back to give proper clearance.
8. The wingwalls of the bridge at 48 chains should be refinished and coped.
9. An undertaking (under sign and seal of the Company) as to the proposed method of working the single line should be furnished.

I must report that in consequence of the incompleteness of the works, the line between South Shields and Marsden cannot be opened to passenger traffic without danger to the public being [*sic*] the same.

I have, etc

C.S. Hutchinson

Major General

The official Harton Coal Company brand.

Harton Coal Company newspaper advertisement of December 1887.

PATENT FUEL

(BRIQUETTES).

THE HARTON COAL COMPANY,

LIMITED,

Beg to inform Consumers that they have Erected EXTENSIVE WORKS and MACHINERY for the Manufacture of

PATENT FUEL FROM THEIR CELEBRATED HOUSEHOLD COALS,

and that they are prepared to supply this ECONOMICAL FUEL (in Large or Small Briquettes) direct to Consumers.

PRICES.	Per ton.
PATENT FUEL WORKS AND COAL DEPOTS, SOUTH SHIELDS	8s 4d.

Carriage extra according to distance.

PRICES.	Per Ton.
To NORTH SHIELDS	10s 10d
„ TYNEMOUTH	11s 3d
„ CULLERCOATS	12s 1d
„ WHITLEY	12s 6d

Including Cartage, and not less than One Ton.

A WEIGHT TICKET SENT WITH EACH LOAD.

For the convenience of Customers the Company will provide, when required, the necessary Porterage, viz. —

Putting into Hatch, 6d per load; Putting into Yard 9d per load.

Orders received at the St. Hilda Colliery, at the Depots, or by letter addressed to

THE HARTON COAL CO., LD., COLLIERY OFFICES, SOUTH SHIELDS.

It is clear from this report that Major General C.S. Hutchinson did not inspect the initial 800 yards of railway from Whitburn Colliery Junction or the final 1,200 yards from Marsden station to Whitburn Colliery but instigated his inspection from the South Shields terminus which was situated just east of Westoe Lane to the other terminus at Marsden.

The Whitburn Coal Co. duly took the Major General's findings on board and after reinspecting the line eight months later, he submitted the following report on 14th February, 1888:

Sir,

I have the honour to report, for the information for the Board of Trade, that in compliance with the instructions contained in your minute of the 1st instant, I have reinspected that portion of the Whitburn Coal Company's Railway which extends from South Shields to Marsden, a distance of 2¾ miles.

I find that the requirements recommended in my report of the 20th June last have been complied with except that the wingwalls of the bridge at 48 chains have not yet been finished and coped though the work is in hand and that a urinal is not yet completed at Marsden station.

Proper signal arrangements have been provided at South Shields and Marsden, the railway's two stations; at South Shields the cabin contains 10 levers and 2 spare ones and at Marsden 12 working ones and 3 spare ones.

The only further requirements witnessed were

1. That the platform should be extended at South Shields station and should end in a ramp.
2. That the facing points at South Shields must be provided by a fang tie.
3. That at Marsden station numbers 9 and 14 levers should be interlocked.
4. That signal diagrams, nameboards, and outside clocks should be provided at each station.
5. That some of the chairs needed additional spikes.

These requirements as well as the two of my report of the 20th June last are to be at once complied with, and upon this understanding and upon receipt of a satisfactory undertaking as to the working of the finished line (this undertaking was handed to me this morning for the secretary's signature) I can recommend the Board of Trade to sanction the opening of the line between South Shields and Marsden for passenger traffic.

I have, etc
C.S. Hutchinson
Major General

Diagrams of the stations showing the signal arrangement should be forwarded to the Board of Trade.

CSH

Thus nine years after the railway first opened, passengers could be officially carried for the first time. In fact the first passenger trains began running on 19th March, 1888 with a timetable catering for six trains on a weekday and seven on a Saturday. A Sunday service was instigated in September 1888 and the earliest surviving timetable, that of January 1890, lists the following service:

This headstone complete with rather touching inscription can be found in St Peter's churchyard, Harton. The stonemason has gone to some trouble to produce an accurate carving of a Whitburn Coal Company engine of the day.

Ian S. Carr

Weekday trains depart South Shields
8.15, 10.00 am, 12.50, 1.00, 3.15, 4.35, 6.15, 9.30 pm

Weekday trains depart Marsden
9.15, 11.55 am, 12.35, 2.45, 4.15, 5.35, 9.00 pm

Saturday trains depart South Shields
8.15, 10.00 am, 12.50, 3.15, 4.35, 6.15, 9.30 pm

Saturday trains depart Marsden
9.15 am, 12.35, 2.45, 4.15, 5.35, 9.00 pm

Sunday trains depart South Shields
2.00, 3.00, 5.00, 6.00, 7.00, 9.30 pm

Sunday trains depart Marsden
1.00, 2.30, 4.30, 5.30, 6.30, 9.00 pm

Each train was allowed a generous journey time of 10 minutes to cover the 2¾ miles between the termini.

The early timetables only listed services for the general public and there were separate trains for the pit and quarry workers, these separate services running beyond Marsden to a private platform at Whitburn Colliery itself. Passenger traffic was bolstered in 1889 when the North Eastern Paper Co. Ltd opened a chemical works just south of Whitburn Colliery. This became a fully fledged paper mill in 1895 and produced some 80 tons of newsprint paper every week which was then distributed to newspaper printers throughout Britain. Although an entirely separate concern from the Whitburn Coal Co., the plant did require 3 to 4 tons of coal for every ton of paper produced and a short rail link ran from the pit sidings directly into the mill to supply the aforementioned coal.

After stuttering into life, Whitburn pit began to produce a fairly respectable tonnage of coal but geological and therefore financial difficulties hampered its true potential. So much so that in 1891 the Harton Coal Co. decided to dissolve both the Whitburn Coal Co. and its sister company the Harton & Hilda Coal Co., and directly run all its South Shields pits as a single empire. Money was injected into the Whitburn enterprise, and the pit's five workings which stretched a mile or so out under the North Sea were soon producing an average of 2,600 tons of coal per day.

Bringing Whitburn Colliery back into 'the Harton fold', was in fact the first stage of a 25 year campaign to turn the Harton Coal Co. into one of the most advanced colliery systems in the country. Lindsay Wood had now taken over from his father as Chairman of the company and was already visualising a unification of his pits by means of an internal railway system complete with private shipping facilities, rather than having to rely on the NER to forward HCC coals to what was then the nearest port, Tyne Dock.

In 1892, Wood purchased an old rundown ballast railway which wound its way up from a quay on the Tyne, past St Hilda Colliery, and up to 'the Bents' (the site of the planned Benthouse pit) terminating just a few hundred yards north of the Marsden Railway. A decade would pass before this ballast line could become part of the Harton system but it was already clear that the Marsden Railway would

This busy Marsden Quarry scene shows the quarry floor interlaced with a network of two foot gauge tracks with no less than four cranes in operation. In the foreground, a horse is hauling three of the iron side-lip wagons mentioned in the text. The tracks were constantly relaid over the years to suit the excavations. The year is 1904. *Public Record Office, Durham/British Coal*

soon become more than simply an isolated branch line. The passenger services which had hitherto started and terminated at South Shields were, from 1895, run from the Marsden end of the operation leaving the South Shields terminus clear of stabled carriage stock. The year 1895 also saw a start made on the complete rebuilding of Whitburn Colliery. The original two-road shed to the north of the pit was demolished and a new one (also with two roads) built to the south. The miner's platform also disappeared from the pit's eastern flank only to reappear on the western side directly alongside the winding towers. The shafts at Whitburn were by this time 184 fathoms deep, and both were equipped with new horizontal winding engines of 48 inches bore and 72 inches stroke, each capable of handling 2,000 tons of coal per day. The Marsden quarries were seen as part of the Whitburn operation, and the two Lighthouse quarries were upgraded, two new brick-built kilns being constructed alongside the battery of seven original kilns.

The early years of the 20th century saw the Harton empire approaching its zenith with a workforce of 10,000 on its books and modernisation continuing apace. However, the relationship between manager and workman was often troubled particularly at Whitburn, a pit which was to become notorious for its militancy over the years, and this problem frequently overspilled to affect the railway staff (normally seen as an entirely separate and somewhat elite workforce). On 9th December, 1899 it was noted that railway staff were being expected to work for 48 hours at a stretch, being allowed just an occasional break to go home and fetch fresh supplies of food, and that some men were having to work 20½ shifts per fortnight! On 27th February, 1900 the men put in an official complaint to the management who immediately responded by putting inexperienced drivers on the locomotives, forbidding the regular railwaymen to eat their food on company property, and by forcing some men to work long shifts and other men, short shifts. Naturally the railwaymen were outraged at being treated so shabbily and although the dispute was eventually resolved, bad feeling remained for several years.

On a happier note, the turn of the century saw the increase in mineral traffic on the railway (647,233 tons of coal were produced by Whitburn in 1900) matched by an upturn in passenger trade, and the passenger service in 1900 ran thus:

Weekday trains depart Marsden
7.25, 8.45 am, 12.20, 4.20, 5.35, 10.05 pm

Weekday trains depart South Shields
8.15, 10.00 am, 1.40, 4.35, 6.00, 10.30 pm

Saturday trains depart Marsden
7.25, 8.45 am, 12.15, 12.35, 3.00, 4.20, 5.35, 9.00 pm

Saturday trains depart South Shields
8.15, 10.00 am, 12.50, 1.40, 3.20, 4.35, 6.00, 9.30 pm

Sunday trains depart Marsden
6.20, 9.00 am, 1.00, 2.30, 4.30, 5.30, 6.30, 9.00, 10.05 pm

Sunday trains depart South Shields
7.10, 10.40 am, 2.00, 3.00, 5.00, 6.00, 7.00, 9.30, 10.30 pm

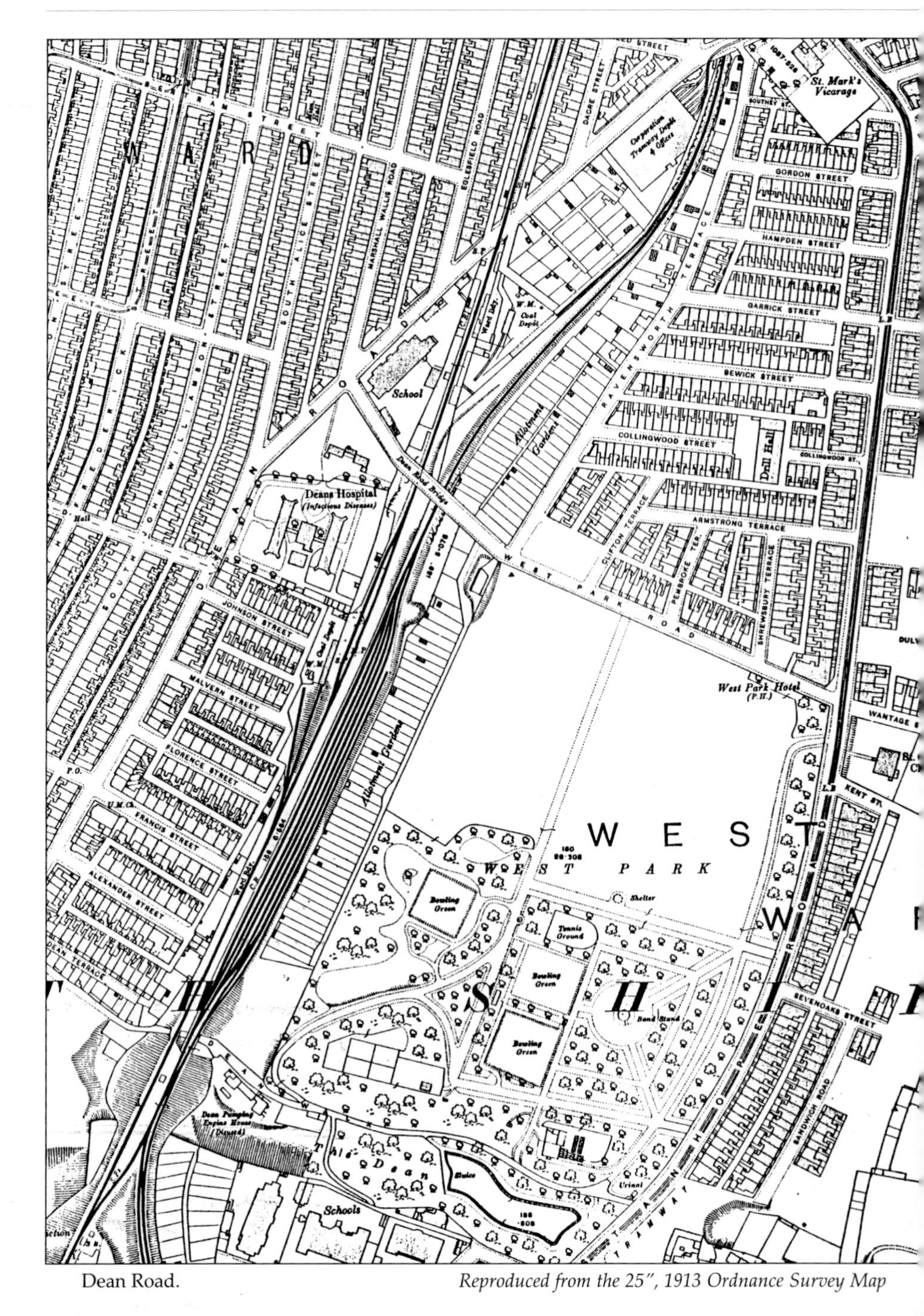

Dean Road.

Reproduced from the 25", 1913 Ordnance Survey Map

As can be seen, weekend passenger trade was brisk due to miners and their families making regular visits into South Shields, and holiday makers making excursions out to Marsden in order to visit Marsden Bay (a local beauty spot which boasted a grotto built into the cliffs, nesting sites of rare seabirds, and the Marsden Rock). The HCC capitalised on this trade by upgrading the railway in 1900. They completely rebuilt the station buildings at South Shields, widened the platform, and renamed the station Westoe Lane (even though it was still referred to as South Shields on the tickets). They also built a little halt, Marsden Cottage, 1 mile and 6 furlongs from Westoe Lane station to cater for the new housing estates at Horsley Hill. They even renamed the actual railway 'the South Shields, Marsden, and Whitburn Colliery Railway' although locally it was still referred to as the Marsden Railway. First class passengers were charged 9*d*. for a single journey to Marsden and 1*s*. 3*d*. for a return ticket, while third class passengers were charged 4*d*. single and 6*d*. return. Although this must have been lucrative for the HCC, it was still seen as something of a sideshow to the mineral traffic and whereas the company invested in brand new wagons where possible, coaching stock was strictly second-hand. In fact by the time coaches were delivered to the railway, they were virtually life expired which the general public found disconcerting to say the least. This quote came from the 23rd April, 1901 edition of the *Shields Morning Mail*:

> A stranger in these parts who had occasion to journey to Marsden the other day has remarked to me on the old fashioned and uncomfortable character of the carriages, and also what he considers the excessive fare charged for such a short journey. He is not the first by any means who has held these views, nor will he be the last. It is a matter for wonder to me that the Company owning the line to that charming spot, Marsden Rock, do not encourage holiday makers to visit it. The matter rests with them.

This inferior 'ride' gained some notoriety and a nickname was given to the passenger service, 'the Marsden Rattler', which remained with the trains until closure. With pressing business elsewhere, the HCC took little heed of the complaints. In 1903 they completely upgraded the former ballast railway in readiness for their mineral trains and linked it up with the Marsden Railway across the few hundred yards of waste ground known as the Bents. By 1904 they had built staiths down on the River Tyne on the Quay frontage at Mill Dam and by 1908 they had upgraded a second old wagonway which ran from St Hilda Colliery to another quay further up river. Here they rebuilt a smaller set of staiths which became known as Harton High Staiths while the set at Mill Dam became known as the Harton Low Staiths. This new internal railway system had a profound effect on mineral traffic flow on the Marsden Railway. Previously all minerals from Whitburn had been hauled down to the NER at Whitburn Colliery Junction, but now the vast majority of this traffic was re-routed across the Bents and down the former ballast railway (which was now known as Erskine Road Bank) to the Low Staiths.

In 1908 the HCC bestowed a crowning glory to its new railway - electrification.* This was by means of electric locomotives working off overhead wire at 550 volts DC supplied by the German company Siemens-Shuckert. The overhead wires ran from both sets of staiths and up the 1 in 38 Erskine Road Bank to the Bents where exchange sidings were installed to allow for the change over from steam to

* See the author's *The Harton Electric Railway* (Oakwood Press - OL91).

Chichester Road. *Reproduced from the 25", 1913 Ordnance Survey Map*

electric traction to take place. Thus the Whitburn locomotives now had to work only as far as the Bents.

Further modernisation came in 1910 when a ½ mile branch was laid from Harton Colliery down to Dean Road just south of Whitburn Colliery Junction. New sidings were laid at Dean Road, Whitburn Colliery Junction was abandoned, and the NER was relinked into the new railway just south of the sidings at what became known as simply Whitburn Junction.

No sooner had this new branch been laid than it became electrified, again via 550 volt DC overhead wires, and the wires ran not just down the Harton Colliery branch, but beyond up the original section of the Marsden Railway (a section which was now known as Chichester Road Bank), through Westoe Lane station and into the exchange sidings at the Bents via a new spur.

Indeed this former section of the Marsden Railway had now disappeared into the new central electrified system and the Marsden Railway itself now officially began at a zero mile post 100 yards west of Westoe Lane station before running the remaining 3½ miles to Whitburn Colliery.

Benthouse (later Westoe) Colliery opened in 1913 on the site of the Bents adjacent to the exchange sidings and although it would be many years before this pit would see full production, it was viewed at the time as the last piece of the 'Harton jigsaw'.

At this stage in history it would be easy to imagine the brand new Harton Railway (complete with its ultra-modern electrification and twin staiths) taking the HCC right into the 20th century, with the Marsden Railway (and its suddenly antiquated 'rattler') left to become a branch line lost in time and kept open simply to 'make the numbers up'. And this would have been true were it not for one simple factor - Whitburn Colliery. This pit was not only producing what was said to be the best steam coal in the North-East, it was producing it in vast quantities. In the first decade of the 20th century half a million tons of coal was produced at Whitburn year in year out. In 1913, the output of machine-got coal (i.e. from mechanical picks, drills, machine cutters, etc.) was 118,596 tons, almost half the total machine-got coal for all the Harton pits, and anyway the colliery did not entirely escape the spread of electrification, with electrically driven power at the pit amounting to 3,150 horse power. The Marsden quarries of course also made a contribution to minerals carried. True, dimension stone had ceased to be quarried at Marsden back in 1908, but only because production was totally committed to screen stone and lime.

Although the Harton Railway had its 10 'state of the art' electric locomotives, the existing steam fleet at Whitburn in 1913 was still healthy and boasted the following examples.

No.	*Type*	*Built by*	*Works No.*	*Year*
4	0-6-0ST	Black, Hawthorn	826	1884
5	0-6-0ST	R. Stephenson & Co.	2629	1887
6	0-6-0	R. Stephenson & Co.	2056	1872
8	0-6-0	R. Stephenson & Co.	1973	1870
9	0-6-0	Blyth & Tyne Railway	-	1862
10	0-6-0	Blyth & Tyne Railway	-	1862
11	0-4-0ST	Manning, Wardle	?	?

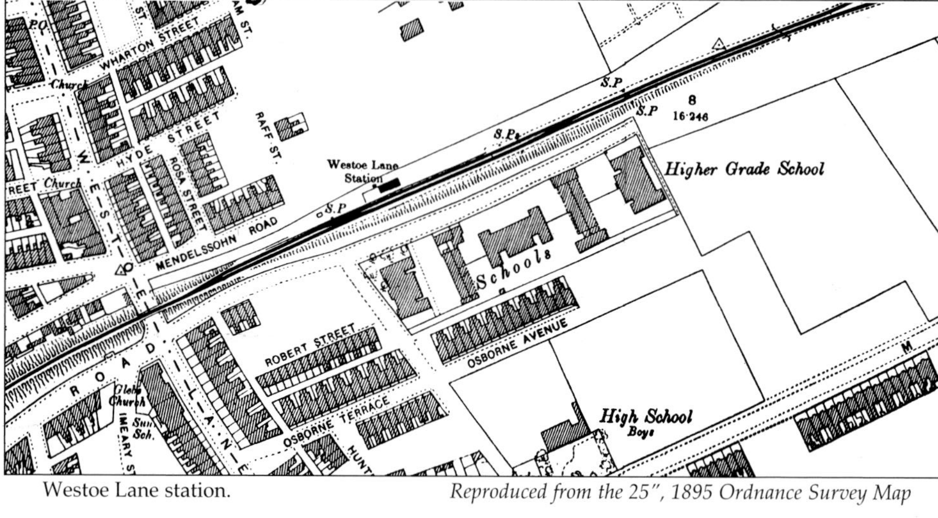

Westoe Lane station. *Reproduced from the 25", 1895 Ordnance Survey Map*

This rather clean looking shot of Westoe Lane station was taken in 1907 a year before the overhead wire paraphernalia appeared. The extensive building indicates just how seriously the HCC once took the passenger service. The signal box can be seen jutting out midway along the platform while the architect seems to have been fanatical about chimneypots! Westoe Colliery lay beyond the footbridge. *Public Record Office, Durham/British Coal*

Chapter Two

A Journey Down to Whitburn

Having described the history of the South Shields, Marsden & Whitburn Colliery Railway up to the year 1913, it would be pertinent here to pause and detail how the railway must have looked at this time before the great changes to the little branch line took place.

A stranger to the railway wishing to travel down to Marsden may have had difficulty in locating Westoe Lane station in South Shields. Westoe Lane itself by 1913 had become Westoe Road, and the station was never situated here anyway but in Mendelssohn Road, a side street. The traveller, having tracked down the station would have been offered a first or third class ticket (even though first class travel on the railway had been abolished at the turn of the century, first class tickets continued to be available for many years afterwards!).

With ticket in hand, entry would have been gained onto the long single platform, with the station itself consisting of offices for station master, wages, tolls, and bookings, workmen's waiting room, public waiting room, station master's house, station staff houses, and a signal box all incorporated into one large, rambling building.

Three lines passed the station, the outer two being equipped with overhead wires, and the traveller may have witnessed an electric locomotive gliding past with a rake of coal trucks in tow, the olive green of the engine contrasting with the brick red of the 10½ ton wagons. A view westward would see the three tracks running for 124 yards before 'disappearing' over the top of a bank. Home signal number 14 stood on the edge of this bank along with a zero mile post which marked the official start of the South Shields, Marsden & Whitburn Colliery Railway, and engines used this short section to run round the passenger trains. On their return from the mile post, the light engines would climb towards the station on the 1 in 51, passing over bridge No. 1 (Imeary Street) and bridge No. 1a (Westoe Lane). Meanwhile miners would be boarding one section of the train while members of the public were being escorted by the guard to another section (for by 1913 each train was being run for workmen and public alike).

By now the black-liveried engine could be seen taking on water from a column situated at the eastern end of the platform (which is why this nearside track was not equipped with overhead wire). The train crews used this column whenever possible as the water here was a good deal 'softer' than that at Marsden.

Once coupled up, the train made a fierce exit from Westoe Lane, starting as it did on a gradient 1 in 185. Beyond the station it passed under bridge No. 5 (Iolanthe Terrace footbridge) where a water tank supplying the station water column could be seen on the right (in the form of a redundant boiler). Westoe cemetery could be seen from the left side of the train with several tracks curving off northwards behind this toward a small collection of pithead buildings almost church-like in design - the fledgling Westoe Colliery. A second set of

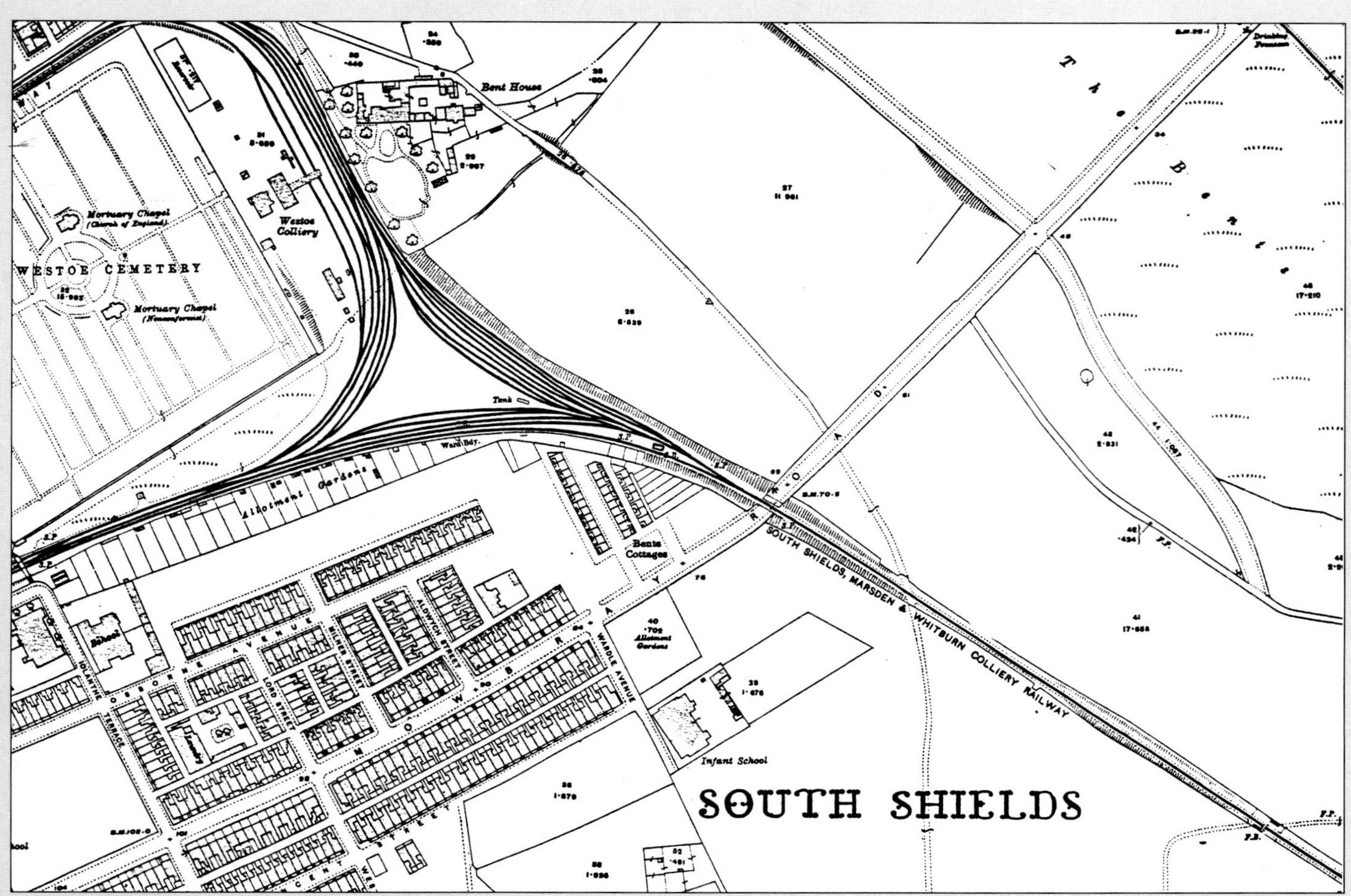

Mowbray Road.

Reproduced from the 25", 1913 Ordnance Survey Map

sidings ran parallel with the left side of the train and these would have been filled with more of the brick red 10½ ton wagons with perhaps a rake of earlier chaldron wagons now serving as a platelayers train. A third set of sidings could be seen sweeping in from the north forming a triangle with the other two sets and, where this third set met the Marsden Railway, there was a signal cabin, Mowbray Road Bridge, perhaps with the olive green electric locomotive standing alongside awaiting release from its wagons. Beyond this the 'Rattler' crossed over bridge No. 2 (Mowbray Road) and in an instant the scenery had changed completely.

Gone was the housing, lineside allotments, sidings, and overhead wire. The train was now out in open countryside running along a single track while to the right could be seen the rugged Cleadon Hills and to the left, across the common, the sea. Benthouse Lane was passed, one of several footpaths which crossed the railway on the level, these all being protected by kissing gates and 'WHISTLE' notices. Still climbing on a gentle 1 in 1329, the train passed under bridge No. 8 (Trow Rocks footbridge) and just beyond this a fixed distant signal and the one mile marker post.

The railway here ran arrow straight and just level with the surrounding fields crossing first Horsley Hill 'Road' and then Little Horsley Hill 'Road' (in truth each a farm track) before the train began to slow for Marsden Cottage Halt. If passengers wished to alight here, having purchased their 2*d*. single tickets, they told the guard at Westoe Lane and he in turn then went up to the driver and simply said 'half way'. The driver then knew that he had to stop here. As the train slowed, it crossed the road to Marsden Cottage which was in fact an isolated hamlet of several houses and these could be seen across the downs from the left of the train. Marsden Cottage was formerly the gatehouse to Salmons Hall, a long derelict house, and strangely, the railway staff persisted in calling the Halt itself Salmons Hall rather than by its official title.

The Halt platform stood to the left of the train and was just long enough to accommodate one carriage while a brick shelter 6 feet long, 4 feet wide and 8 feet high offered refuge to any waiting passenger. Immediately beyond the Halt, the train crossed a footpath which climbed off to the right towards Harton Down Hill, an area of high ground pock-marked with old quarry workings. Once on its way again, the train soon passed the 2 mile marker post and crossed over bridge No. 3 (Redwell Lane). At the time, Redwell Lane was little more than a footpath which lead down to Marsden Bay and the clearance underneath the railway bridge was just 6 ft 1 in.

Beyond Redwell Lane, the coastline swept right up to the edge of the railway and just a few yards from the clifftop, the train crossed a footpath which ran between two public houses, Marsden Inn and Marine Grotto. Marsden Inn could be glimpsed off to the right but the Grotto was hidden away under the cliffs to the left being accessible only by some steep steps.

Shortly after this, sidings would have suddenly appeared, running alongside the right flank of the train, and then a trailing spur, curving away inland before disappearing under Lizard Lane. This was the by now virtually disused branch to Marsden Old Quarry and, beyond Lizard Lane, it ran right into the quarry where just one set of points allowed for some limited shunting.

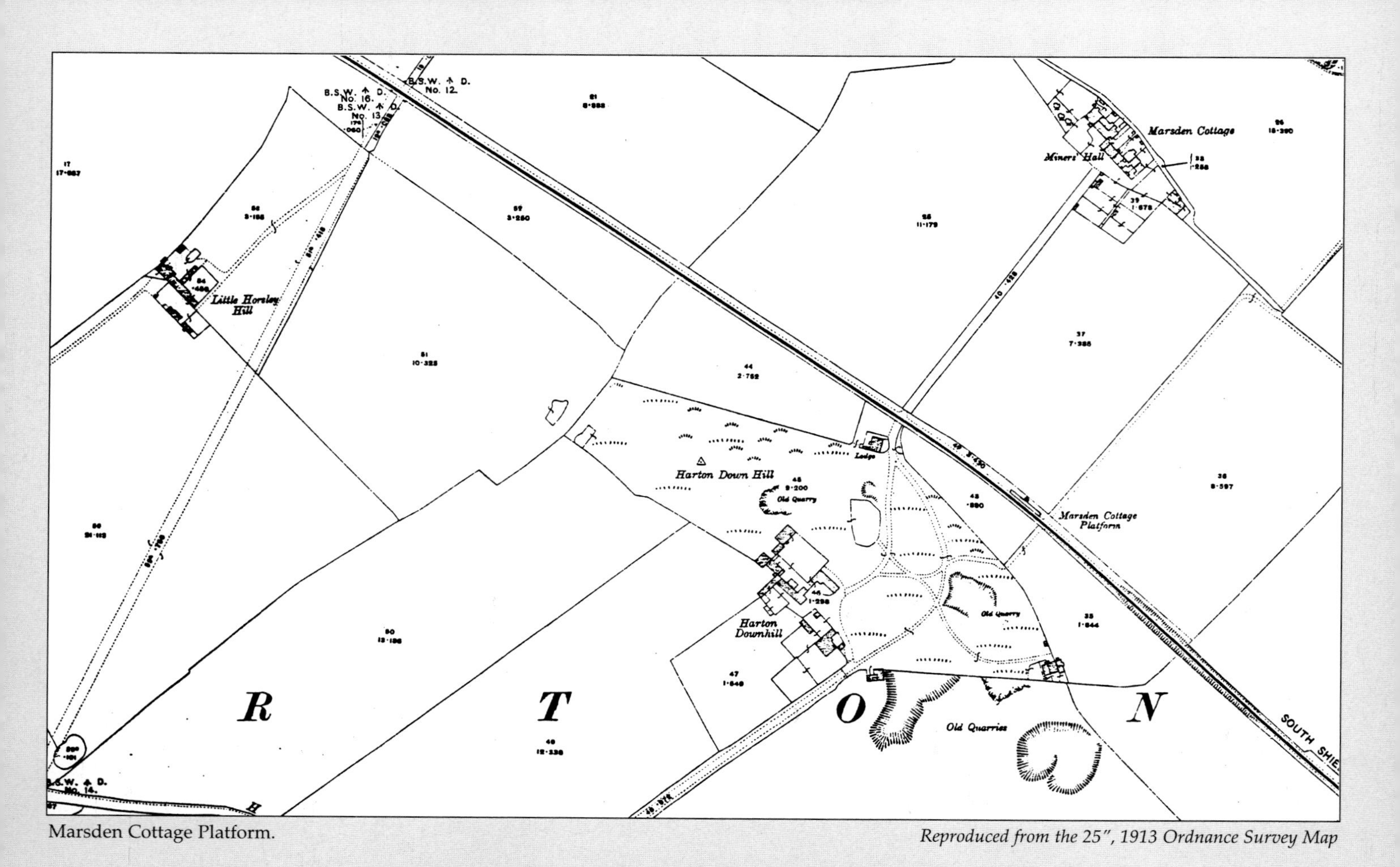

Marsden Cottage Platform.

Reproduced from the 25", 1913 Ordnance Survey Map

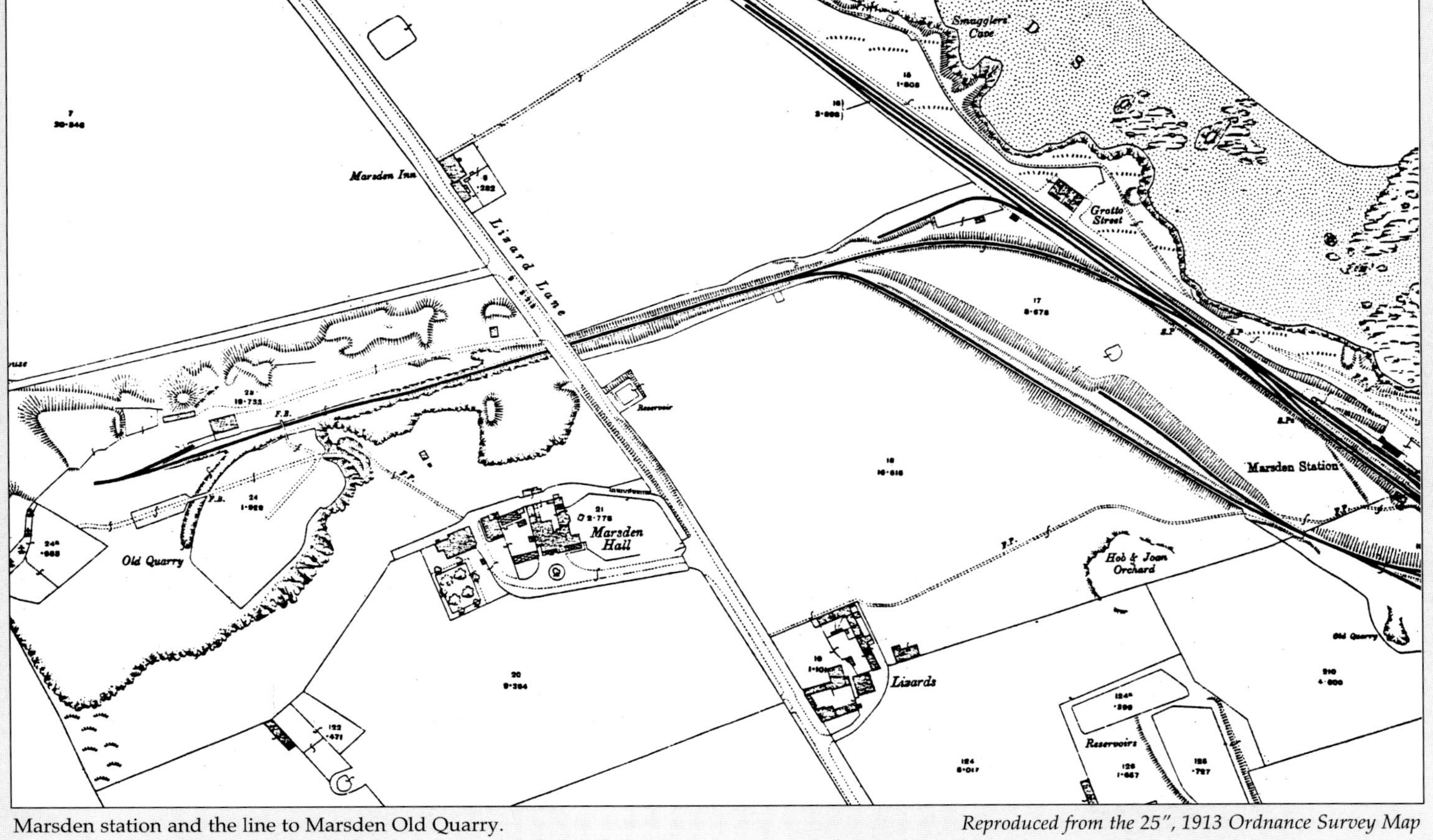

Marsden station and the line to Marsden Old Quarry. *Reproduced from the 25", 1913 Ordnance Survey Map*

This is how Redwell Lane looked when it really was a lane and not the road it is today. This rather rickety-looking bridge gave road traffic an overhead clearance of just 6 ft 1 in. and was heavily rebuilt in 1937. The pedestrians are heading toward the coast road. *Shields Gazette*

This 1904 photograph illustrates the Marsden limekilns viewed from the south, with the original Marsden station in the distance. The Marsden Railway main line passes to the right of the 10½ ton wagon while a further 10½ ton example can be seen positioned on one of the low-level tracks ready for loading inside the kiln shed. The high-level sidings serving the kiln vents are hidden behind the heavy retaining wall to the left. *Public Record Office, Durham/British Coal*

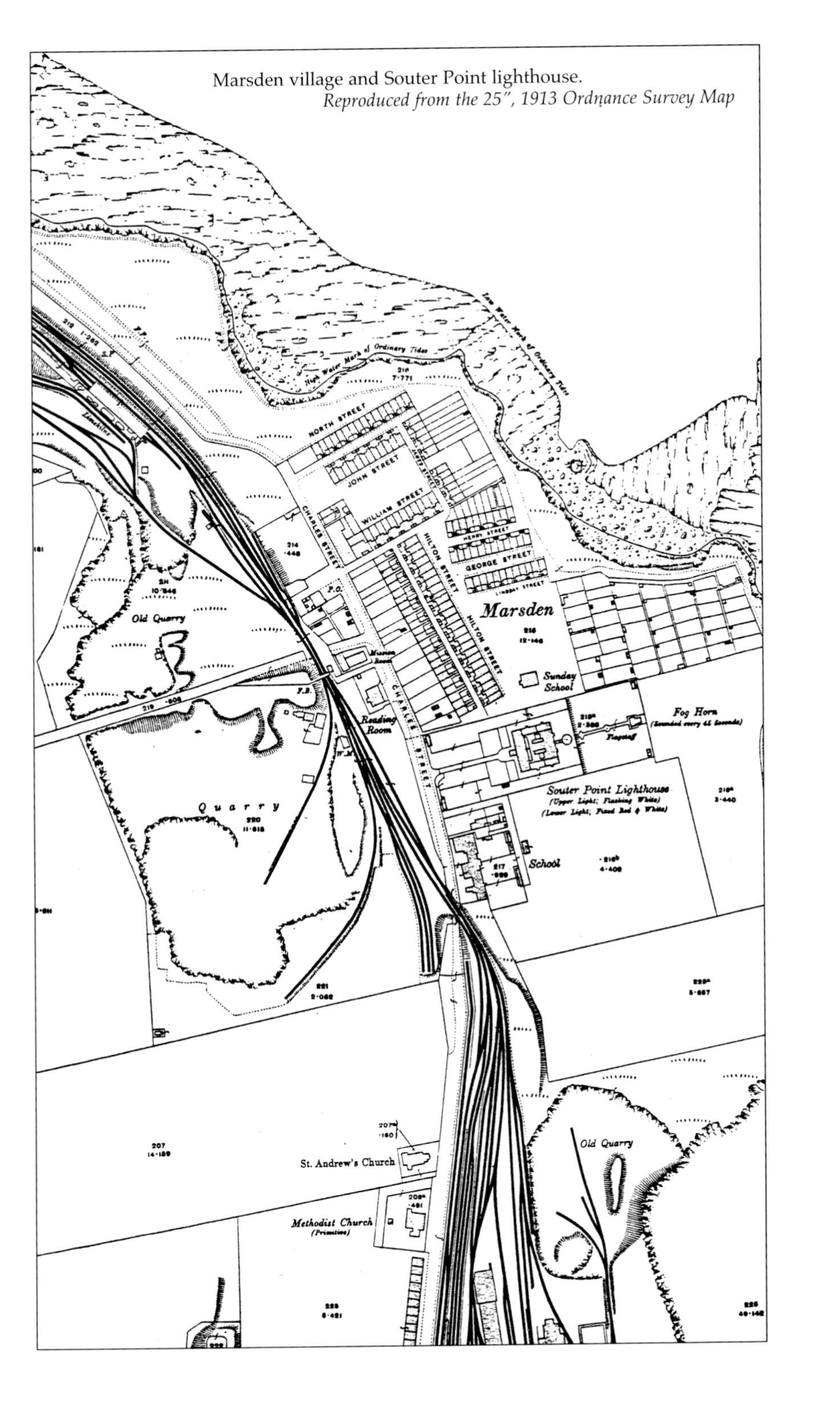

Marsden village and Souter Point lighthouse.
Reproduced from the 25", 1913 Ordnance Survey Map

With the sidings still alongside, the train came to a halt at Marsden station. In keeping with the design at Westoe Lane, Marsden had just one platform (to the left of the train) water column complete with 'boiler' water tank (on the opposite side of the railway) and a ground frame rather than a signal box built on the platform itself, though the rest of the station buildings were much more limited than those at Westoe Lane. The station was in fact served by a footpath which zig-zagged down to a lane, which in turn ran off to Marsden village.

The tracks running through the station were on three levels. The platform track and run-round loop were on mid-level while immediately beyond, the sidings had dropped down to a low level. One hundred yards beyond this stood a high level track which was in fact a second spur running from Marsden Old Quarry.

With all members of the public safely detrained, the train could continue and as it left the station, the land became dinstinctly industrial again with the three level tracks all running parallel before the Marsden lime kilns were reached. Here the low level sidings ran underneath the kiln while the high level sidings swept in to run directly above, with the 'Rattler' continuing on its course on the mid-level track to the seaward side of these.

There was a battery of seven kilns with the two newer brick-constructed kilns built on to the southern end. Just two men worked the top of the kilns, opening the hopper doors of the wagons, which were standing on the high level sidings, and allowing the minerals to cascade into the vents in ratios of 4 tons of limestone to 1 ton of coal. Then, in what was a continual process, two men raked the burned lime out through the draw arches at the bases of the kilns and allowed it to cool before barrowing it into further wagons standing in the low-level sidings beneath.

Once the train had passed the kilns, Marsden village came into view on the left-hand side with its little rows of back-to-back terraces while a siding also to the left of the train served Charles Street Landsale Coal Depot. The depot yard was situated below the siding and once the hopper doors of the wagons were opened, the coal teemed down between the tracks into the low-level cells where it could be bagged up according to size and grade, and delivered by horse and cart to houses and local business alike.

Close to the third mile post, the train passed under bridge No. 6 (Mission Room footbridge) while each side of this footbridge two opposing spurs ran off to serve the two Lighthouse quarries which were for the most part hidden away from the main line. Inside the quarries the little 2 feet gauge tracks ran from the working faces down to wooden gantries constructed directly above the standard gauge spurs. Here the wagon loads of limestone were side-tipped into the standard gauge rakes of trucks prior to removal. One steam locomotive shunted the standard gauge spurs while 33 horses and 104 iron side-tipped wagons were employed on the narrow gauge lines.

Immediately beyond the village, and dominating the skyline to the left of the train was Souter Point Lighthouse, 76 feet high and equipped with a foghorn which was powered by stationary steam engines. The sports field alongside the foghorn was home to Marsden Welfare football team, a club which enjoyed a remarkably successful home record due to an unusual coaching technique. If

Level crossing gates on the Marsden Railway were rather basic affairs compared to their main line counterparts and lacked both the extensive use of white paint and the red warning discs. This particular example guarded the lane which ran from Lizard Lane to The Coast Road. In fact the coast road can be seen beyond the crossing as can Marsden village with the Mission Room footbridge to the right. The year is again,1904.

Public Record Office, Durham/British Coal

the team looked like losing a match, the coach would rush over to the lighthouse and instruct the keepers to test the foghorns out. It was said that the noise was unbelievable and that the ground used to shake. The home team used to practice on the field and had become accustomed to the din, but for the away players, the effect was totally demoralising. After half an hour or so the away managers would ask for matches to be abandoned with automatic wins awarded to the home side!

Continuing its journey, the train passed Marsden school on its left and crossed over bridge No. 4 (Mill Lane). The only bridge of any substance since Mowbray Road, Mill Lane marked the boundary between the Marsden quarries and Whitburn Colliery. Two tracks crossed the bridge with all mineral traffic using the seaward track while the 'Rattler' exclusively used the track to the landward side. Beyond the bridge, extensive sidings fanned out to the left of the train while to the right, a single lengthy spur acted as a carriage siding for any spare coaching stock. Here the train slowed toward the end of its journey, Whitburn Colliery station, which again in keeping with the other stations on the railway, consisted of a single platform on the seaward side of the line but this time with no station buildings, water column, or signal box. There was a loop here to allow engines to run round and the SSMWCR officially ended in a short headshunt in front of the colliery offices and butting up to the main pit entrance.

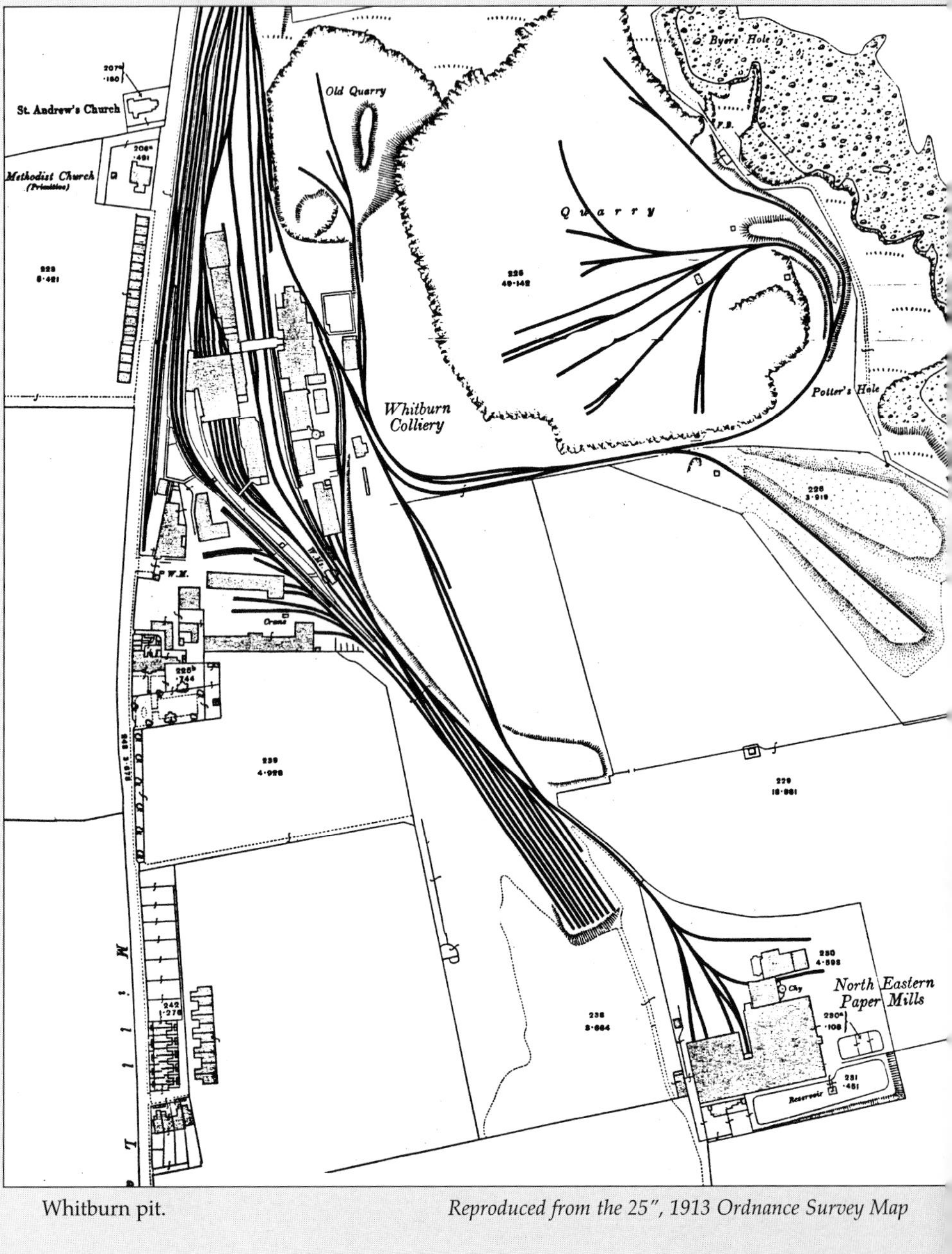

Whitburn pit.

Reproduced from the 25", 1913 Ordnance Survey Map

Here the miners detrained below the pit's two winders (the number one winder was a steel structure, while number two consisted of steel, encased in concrete) and the 14-year-old apprentices travelling to work for their very first experience 'down pit' must have looked up at the accelerating winding pulleys spinning into a blur with some trepidation.

The sidings, of course, continued beyond the SSMWCR in order to serve the pit. The bulk of these sidings, known as the bank sidings, ran beneath the screens to facilitate wagon loading while a secondary spur ran round the seaward side of the pit and into the harbour quarries for disposal of colliery spoil. Immediately south, and adjacent to the main pit buildings, was Whitburn shed which was served by two roads coming into the shed from the south. A further siding flanked each side of the shed, and locomotives were often seen here either stabled or taking on coal and water. To the west of the shed the bank sidings were extended to engage with a pair of weighbridges before running down to storage sidings beyond the colliery. Further sidings served a yard inside the main entrance which was home to the colliery stores, workshops, and a landsale coal depot. A final spur, south of the storage sidings, served the North Eastern paper mills some 500 yards beyond Whitburn Colliery.

This panoramic photograph of recently rebuilt Whitburn Colliery was taken *circa* 1900 and shows the little Black, Hawthorn locomotive No. 2 about to leave the original miner's platform with a train of early ex-NER stock for South Shields. Meanwhile, No. 10, one of the ex-Blyth and Tyne locomotives, is at the head of a lengthy train of 10½ ton mineral wagons. The pit boilerhouse is directly behind the passenger train.

Beamish, North of England Open Air Museum

This official photograph of Whitburn Colliery taken in the 1920s shows the pit offices with the SSMWCR terminating in front. The station platform can be seen off to the left and entry to this was gained through the metal gates next to the bufferstop. Miraculously, the two massive concrete entrance pillars on the right survive to this day. *Fred Bond Collection*

One of the 'ornamental' railway bridges is seen straddling Westoe Road in this 1925 scene. The zero milepost which marked the commencement of the SSMWCR was situated next to signal No. 14 seen on the extreme right with HCC overhead wire support posts alongside. Westoe Lane station was situated in a side street which ran off to the left just in front of the bridge. A Corporation tramway support post dominates the scene. *South Tyneside Libraries*

Chapter Three

The Rise and Fall of the 'Rattler'

The Marsden Railway settled into something of a routine after the tumultuous events leading up to the year 1913. Coal production at Whitburn was impressive with 575,899 tons produced from 262 working days in this year. Excavated tonnages were also high with the Lighthouse quarries yielding 89,079 tons of limestone and 15,209 tons of lime.

The market was buoyant with the HCC charging 11 shillings per ton for exported coal (a price incidentally, which infuriated local residents who at the time were being charged 18 shillings per ton for landsale coal!). Profits made by Whitburn alone for the year 1913 were estimated to be £432,000 while one of the main outgoings was money paid for wayleaves. Here, the Ecclesiastical Commissioners were paid a set rent of £20 per annum plus 1½*d*. for every ton carried on the railway.

Alongside such figures, receipts from the 'Marsden Rattler' looked rather meagre with just £264 taken for the entire year. However, such takings more than paid for the running costs of the train even though the HCC had reduced first class fares from 9*d*. to 6*d*. for a single journey, and from 1*s*. 3*d*. to 9*d*., return. A range of tickets on sale in 1913 is listed below.

Ticket	*Price*	*Type*	*Colour*
First class adult single	6*d*.	card	white
First class adult return	9*d*.	card	white
First class child single	3*d*.	card	white/pink
First class child return	4½*d*.	card	white/pink
Third class adult single	4*d*.	card	white
Third class adult return	6*d*.	card	green/white
Third class child single	2*d*.	card	white/pink
Third class child return	3*d*.	card	green/white/pink
Single to Marsden Cottage	2*d*.	paper	mauve
Workman's weekly pass	1*s*. 6*d*. for six	card	mauve
School child's ticket	7½*d*. for five	card	orange/white
Special meeting ticket	3*d*.	card	green/white

The above list is by no means exhaustive as there were also tickets issued for dogs and prams. Single tickets only were issued for Marsden Cottage Halt and these seem to be at a set price regardless of class. Sunday meetings were held at Miners Halls at Whitburn or South Shields and the special meeting tickets were issued in conjunction with these at 3*d*. for a single journey. Eventually the workman's weekly pass tickets were replaced by brass tokens retained by the miners.

By 1913, each train service was run for both miners and public alike although the two were kept apart in separate carriages. A timetable issued in this year showed a service geared to shift changes at Whitburn Colliery (a trend which was to become more and more apparent in the ensuing years).

Passenger Timetable for 1913

Weekday trains depart Marsden
5.50, 6.45, 8.40, 11.50 am, 12.10, 1.35, 2.45, 4.20, 5.10, 9.30, 10.20 pm

Weekday trains depart South Shields
6.15, 8.15, 9.00 am, 12.20, 1.50, 3.00, 3.20, 4.45, 5.45, 10.05, 11.00 pm

Baff Saturday trains depart Marsden
5.50, 6.45, 8.45 am, 12.10, 12.35, 1.10, 2.05, 3.00, 4.20, 5.15, 9.00, 10.30 pm

Baff Saturday trains depart South Shields
6.05, 8.15, 9.00 am, 12.20, 12.50, 1.40, 2.20, 3.20, 4.45, 5.45, 9.30, 11.00 pm

Pay Saturday trains depart Marsden
6.10, 6.45, 8.40 am, 12.10, 1.10, 2.05, 3.00, 4.20, 5.15, 9.00, 10.30 pm

Pay Saturday trains depart South Shields
6.30, 8.15, 9.00 am, 12.20, 1.40, 2.20, 3.20, 4.45, 5.45, 9.30, 11.00 pm

Sunday trains depart Marsden
5.20, 6.15, 9.45 am, 1.00, 2.30, 4.30, 5.25, 9.00, 10.15 pm

Sunday trains depart South Shields
5.35, 7.10, 10.40 am, 2.00, 3.00, 5.00, 6.00, 9.30, 10.05 pm

The terms 'Baff' and 'Pay' refer to a system of working whereby the pit was open on alternate or 'Baff' Saturdays while the miners were paid fortnightly on the Saturdays that the pit was shut, hence the term 'Pay' Saturday.

The outbreak of World War I had surprisingly little effect on the railway with Whitburn Colliery producing a healthy 2,392,232 tons of coal for the years 1914 to 1918 bolstered no doubt by a war bonus paid to the miners of 2 shillings a week. There was concern that the German navy might target the pit and its little railway so to help allay such fears, an armoured train was constructed at the Whitburn Colliery workshops. Scant details concerning this train survive, however what is known is that the train consisted of several wagons hauled by No. 11 a Manning, Wardle 0-4-0 saddle tank with outside cylinders, and that the train was built to provide training for the local territorial forces.

In 1920, a new depot was built to cater for the electric fleet (which by now amounted to 10 locomotives) within the triangle of lines to the east of Westoe Colliery. New tracks fed the depot from the south, and these were linked up to the Marsden Railway just north of Westoe Lane station which in turn naturally brought about an increase in electric movements through the station.

On 26th June, 1923, a statement was read to a meeting of the South Shields Corporation, a statement which was to change the entire face of the Marsden Railway: 'Resolved - that a deputation consisting of the Chairman, the Vice-Chairman, Councillors Pearson and Young, interview the Rural District Council on the definite project of constructing a new coast road running southward from the Trow Rocks'.

At this time there was no direct road between South Shields and Whitburn and, with road traffic on the increase, the Corporation had decided to resolve the matter. Because this new road was to run through both Corporation and Rural District areas, a joint committee had to be formed.

An engineer's report was submitted to the committee on 30th October, 1923 and it was clear from this that the biggest obstacle to the new road was going to be the Marsden Railway. The railway ran close to the cliff edge at Marsden bay and the engineers suggested buttressing the clifftop to support the road rather than diverting the railway to provide clearance stating that 'The cost of buttressing the cliffs will be much less than the removal of the railway to a more westerly site, a work of great magnitude if not impossible'.

Rightly or wrongly, the joint committee decided to ignore the report and on 17th December, 1923 contacted the Harton Coal Co. to discuss their intention to divert the offending section of the railway. The committee offered compensation costs of £10,000 to the HCC stating that this money must be returned in the event of the railway being taken over by the London & North Eastern Railway (LNER). The committee had assumed, wrongly, that the Marsden Railway would be swallowed up by the LNER as part of the 1923 amalgamation of the big four railway companies. On 3rd March, 1924 the HCC wrote to the committee stating that they had decided to close Marsden station to make way for the new road and build a replacement passenger terminus at Whitburn Colliery. This would '. . . meet a desire which has long been expressed by the passengers using the railway, and by the inhabitants of the district'.

Terms of agreement were drawn up between the Harton Coal Co. and the committee whereby the HCC agreed to divert the railway, while the committee agreed to pay compensation and to pay for any alterations to the line including the new station at Whitburn Colliery. The committee also agreed not to hinder or interrupt the traffic of the coal company along the railway.

All ground work concerning the realignment of the railway was to be under the responsibility of the committee, while the HCC was to be responsible for all rails, point work, and electrical work.

It was clear that a much more substantial railway bridge was required at Mill Lane to allow the new road to pass underneath and that this bridge alone would cost £7,643 15*s*. 9*d*.

Work commenced on the coast road on 30th June, 1924 with the ordering of 6, 100 tons of clinker ash from the HCC with Boldon Colliery providing 120 tons per week (at 3 shillings per ton), and Whitburn Colliery providing 100 tons per week (at 1 shilling per ton). The committee also put in an order for 1,500 tons of limestone from Alston Colliery in Cumbria (an extraordinary decision given the proximity of the Lighthouse quarries!) and the LNER delivered this to Dean Road sidings via an agreement with the HCC.

While the coast road was under construction, traffic along the adjacent railway was minimal due to two long running disputes at Whitburn. The first dispute took place in 1925 and was due to internal problems at the pit. The pit itself closed down from 23rd April to 28th September with only 140 days worked for the entire year compared to the usual 250. Worse was to come. In 1926 the General Strike took place with Whitburn shut down from 1st May to

6th December and here just 103 days were worked and 237,264 tons of coal produced for the year. Events got particularly ugly in 1926 when miners from the St Hilda pit assembled in the street outside Westoe Lane station to prevent Westoe's workforce from getting to work. The Chief Constable called upon mounted police to prevent further trouble and the Riot Act was read to the striking miners. Sadly, several of these miners were injured later during a baton charge by the mounted police.

The Coast Road Joint Committee had problems of their own in 1926. While construction of the coast road was in full flow, the committee suddenly decided to 'main' the road (in other words upgrade its status to that of a main trunk route between South Shields and Sunderland). This in turn meant that a much more substantial bridge would be required to carry the railway over the former Mill Lane. New plans for the bridge work were released on 26th July, 1926 with an estimate submitted by Orr, Watt and Co. Ltd for £4,530 for steel and another estimate by John Lant of Newcastle of £17,916 18*s*. 9*d*. for bridge abutments. These two estimates plus the money required for lowering the road level underneath, brought the final cost of the new bridge to £27,000 (nearly £20,000 more than the original estimate).

In 1926 the Harton Coal Co. applied for a Light Railway Order so that it could run its new route officially as a Light Railway when it finally re-opened. This was granted under Statutory Rules and Orders (1926) No. 1066 thus:

> The South Shields Marsden and Whitburn Colliery Light Railway Order 1926, dated August 7th 1926, made by the Minister of Transport, authorising the deviation, reconstruction, and working of an existing railway in the County Borough and Rural District of South Shields in the County of Durham as a light railway under Light Railways Act 1896 (59 and 60 Vict. c. 48) and 1912 (2 and 3 Geo. 5 c. 19) as amended by the Railways Act 1921 (11 and 12 Geo. 5 c. 55).

Light Railway Order number 1066 then described the existing railway thus:

> The railway is known as the South Shields Marsden and Whitburn Colliery Railway and is 3 miles 3 furlongs and 5 chains or thereabouts in length partly in the Borough of South Shields and partly in the Rural District of South Shields. The railway commences in the Borough at the signalpost 14 feet or thereabouts west of the bridge over Imeary Street proceeding thence in an easterly and south-easterly direction to and terminating at the existing platform at Whitburn Colliery 12 chains or thereabouts south of the Methodist Church in Mill Lane.

The Order then goes on to list the standard rules and regulations associated with all such Orders including the mandatory speed limit of 25 mph (which actually represented an increase in permitted speed from the previous 20 mph laid down by the HCC). Also listed were the following dimensions for the road bridges on the railway:

Bridge No.	*Name*	*Parish*	*Span*	*Height*
1	Imeary Street	South Shields	50 ft	17 ft 1 in.
1a	Westoe Lane	South Shields	46 ft 3 in.	16 ft 4 in.
2	Mowbray Road	South Shields	39 ft 0 in.	15 ft 3 in.
3	Redwell Lane	Whitburn	9 ft 10 in.	6 ft 1 in.
4	Mill Lane	Whitburn	50 ft	16 ft 6 in.

Order No. 1066 was then duly signed by the assistant Secretary to the Minister of Transport, Mr H.H. Piggot.

The year 1927 proved to be a bleak one indeed for the South Shields Corporation. Their decision to 'main' the coast road and the resultant increase in expenditure this consumed actually bankrupted the Corporation so they had little choice other than to close down the roadworks and this they did on 10th June, 1927 with the works only recommencing after a loan was granted from the Government Treasury on 16th April, 1928! Work commenced on the new bridge over Mill Lane, which was officially titled No. 4 (Lighthouse Bridge), on 12th November, 1928, this being a Pratt truss-type girder through bridge of 110 ft span carrying double track, and representing by far the single largest bridge on the railway.

On 9th April, 1929 the railway diversion was finally completed with all railway traffic diverted on this date.

Trains still ran along the original course of the railway from South Shields with the new coast road now running parallel and to the seaward side of the line. As the Grotto was approached, the line deviated gently inland some 50 feet from its original course, passing under a new footbridge which replaced an original footpath crossing. The ground rose quite steeply here necessitating the excavation of a cutting. The Grotto cutting as it became known, was 700 yards long, 80 feet wide and 18 feet deep requiring the removal of 35,000 cubic yards of material. The line then gently deviated back onto its original course just before the lime kilns were reached with the total deviation running for just 1,000 yards. Beyond the kilns, the trains passed a new signal cabin, Lighthouse box (which had been constructed to replace the recently demolished frame at Marsden station). All mineral traffic was routed to the seaward side of the box while the 'Rattler' exclusively used the single track to the landward side. Beyond this, the summit of the Marsden Railway was reached (close to the 3 mile marker post) while further beyond, the trains drifted down towards Whitburn Colliery. Here they passed over the double-track Lighthouse bridge where again mineral traffic used the track on the seaward side, with the main line running on the inland side which allowed passenger trains to reach the new station at Whitburn Colliery.

The coast road which had eventually cost the Corporation £151,585 finally opened on 2nd November, 1929. A small portion of this cost provided the HCC with its new passenger terminus at Whitburn Colliery. This was inspected by Mr E.P. Anderson on 28th March, 1929 who then submitted the following report on 3rd April, 1929.

Sir,

I have the honour to report, for the information of the Minister of Transport that, in accordance with the minute of the 25th March, 1929, I inspected the extension to the platform at Marsden Station on the South Shields Marsden and Whitburn Colliery Light Railway on the 28th March, 1929. This will be brought into use on or after the 8th instant.

The extension, which is shown in the plan accompanying the Company's letter of 7th March, 1929, is 40 feet in length, with a minimum width of 6 feet in the direction of Whitburn Colliery. The platform is of timber, backed and surfaced with earth. The Company's representative undertook to have the earth surface, which is settled slightly levelled up. The lighting consists of a double electric lamp on a post at the north-west end of the platform extension, and another at the entrance to the old platform, which remains in use. The extension is adequately fenced with a wooden fence similar to that of the existing platform, and the end is properly ramped.

Owing to bus competition, the number of passengers using this line, other than the Company's employees who do not use this station, is so small that only one vehicle is being run on each train for them. The length of the platform extension is therefore sufficient. It is necessitated by the realignment of the track required to make room for the coast road.

The extension has been constructed in a satisfactory manner and I recommend that approval be given to its use as soon as the line has been slewed to its new position, completion of which should be reported by the Company in due course.

I have the honour to be,
Sir,
Your obedient servant,
E.P. Anderson

It is clear from Mr Anderson's Report that the 'terminus' was in fact an extension of the former miner's platform and that station facilities simply consisted of the said platform and a small ticket booth. This new station must have caused a little confusion amongst visitors travelling on the 'Rattler'. Having seen the station advertised as Marsden on both ticket and timetable, such members of the public would have arrived at their destination only to find the station nameboards proudly bearing the name Whitburn Colliery! In fact this strange situation continued until the cessation of the passenger services.

A new timetable issued in conjunction with the Light Railway offered the following service (note that the 'Baff-Pay' system had been dropped by 1930).

Passenger Timetable for 1930

Weekday trains depart Marsden
6.20, 7.20, 10.10 am, 12.15, 1.00, 1.35, 2.40, 4.45, 5.25, 9.10 pm

Weekday trains depart South Shields
6.05, 6.35, 9.05, 9.50, 10.50 am, 12.30, 2.15, 4.20, 5.05, 8.40, 10.15 pm

Saturday trains depart Marsden
5.15, 7.20, 11.30 am, 1.15, 4.20, 6.00, 9.15 pm

Saturday trains depart South Shields
5.30 am, 12.35, 1.30, 5.30, 8.40, 9.30 pm

Sunday trains depart Marsden
6.05, 9.45 am, 12.10, 2.10, 4.30, 6.10, 10.00 pm

Sunday trains depart South Shields
7.10, 10.15 am, 12.30, 1.30, 3.40, 5.30, 8.40, 9.30 pm

The 'Rattler' now ran 3¼ miles to the new terminus compared to 2¾ miles to the old one, so the train was allowed 12 minutes rather than 10 minutes to complete the journey thus keeping the average speed at 16½ miles per hour. As mentioned in Mr Anderson's Report for the Minister of Transport, passenger traffic by this time was minimal due to the introduction of the Corporation bus service which was now making good use of the coast road, and the weekend timetables reflect this. At one time a going concern, the 'Rattler' was turning into little more than a rather quaint local curiosity and as time went by, the railway staff took on a somewhat casual attitude where the public service was concerned.

Passengers could purchase tickets from either the ticket office or directly from the guard. However when the guard sold tickets, nine times out of ten he pocketed the money himself, while one regular passenger normally handed in IOUs instead of money! A gang of youths often travelled up to South Shields and back on a Saturday, at Whitburn Colliery they climbed onto the carriage footboards on the opposite side to the platform while the train was waiting to leave. The guard would spy them from the ducket window of his compartment and then turn his back on them which was their cue to climb into the carriage for their free ride! Other regular Saturday passengers included mothers with prams. They found great difficulty getting their prams on the Corporation buses whereas on the 'Rattler' they could simply wheel them into the guard's compartment.

The last bus for Marsden left South Shields at 10.00 pm on a weekday and many an off-duty miner used the 10.15 pm train service instead after enjoying a last minute drink at the public house. One driver recalls how his guard used to wait until the said miners had almost reached the platform before blowing his whistle and leaving the poor men stranded. He said, 'You used to hear the heavy boots clumping up the hill and the laboured breathing, and know that as the noise got closer, that whistle would blow!'

When not offering these rather unique services for regular users, the passenger carriages ran virtually empty, accept for just one day a year, Good Friday. Between the wars, this was a big date in the South Shields social calendar with some 16,000 children attending church services in the town. After the services there was a great march to the market place lead by the marvellous colliery bands. Bands such as the St Hilda Colliery Band (four times world champions) and the Marsden Colliery prize band. The children were each given an orange for marching and the fairground was opened in the afternoon on the sea front. The Harton Coal Co. provided two trains on the Good Friday, one up to South Shields in the morning and the other returning to Whitburn Colliery in the evening. On the eve of Good Friday, all the carriages were cleaned, the locomotive polished, and the train crews given new uniforms.

Two packed trains on just one day a year was not enough to make the service run at a profit of course, and at face value it seemed inevitable that the 'Rattler' would be discontinued once the coast road had opened. True, the train was regarded as something of a last resort by the public, but to the miners and quarry men it was vital. A total of 3,166 men worked the pit and quarries around Marsden and many of these men lived in South Shields and relied on the 'Rattler' as their sole means of getting to and from work. When the train ran no-one noticed it. If for any reason it did not run, then there was chaos. One evening a locomotive was due to take its usual booked load of 36 full 10½ ton

Westoe Lane station 'under the wires' in 1934. Pride of the fleet No. 8 is about to depart with the 'Rattler' while the 11 o'clock goods train has been safely deposited in the Wallside Road on the extreme left. The original pre-1900 platform runs off to the right behind the water column and signalling. *H.C. Casserley*

The NER 'C' class engine No. 8 simmers at Whitburn Colliery platform on 1st April, 1934 prior to taking this 'Rattler' service up to Westoe Lane. An ex-Great Eastern Railway four-wheeled brake nestles behind the locomotive followed by a rake of ex-North British Railway six-wheeled stock. The miner's footbridge can just be discerned in the background. *H.C. Casserley*

wagons from under the screens at Whitburn and up the coast to Westoe. This particular evening there were only 47 wagons in total under the screens so the crew decided to take all 47 in one train in order to save time. Much sand was used to get the train moving on the rising gradient out of Whitburn Colliery. The crew however managed to get the train into its stride beyond the summit only to find it sliding to a halt on the wet rails in the Grotto cutting with the locomotive's reserves of sand exhausted.

After several attempts at a restart, the crew decided to split the train into two sections, run the first section to Westoe, and then return to the cutting to collect the second. Meanwhile the line was blocked of course and miners were assembling at Westoe Lane ready to be taken to work. The Colliery Manager at Whitburn was immediately notified and an attempt was made to find alternative means of transport. Local bus companies were contacted but could not lay-on such transport at short notice, so as a last resort a local taxi service was hired, with the taxis ferrying their bulging cargoes of miners down the coast road to the pit. This little escapade cost the HCC a small fortune and the train crew were summoned to a management hearing. In fact, the crew were convinced that they would be sacked but were eventually fined instead.

On 25th January, 1930 the miners were offered the luxury of pithead baths, built (at a cost of £25,000) directly opposite Whitburn Colliery station. A concrete footbridge connected the baths with the station platform to save miners crossing both the railway and the coast road, while the general public still made their way to the southern end of the platform to make their exit from the station. The baths also brought another bonus to the miners, cleaner trains with the men now able to travel home in fresh, rather than working clothes.

At the end of each shift, after showering, the men assembled in a canteen in the baths, then, when the 'Rattler' was due to leave, the station master rang an electric bell and the men came trooping over the bridge and boarded the train.

As the train approached Westoe Lane station, the carriage doors were already opening with men jumping onto the platform from the slowing coaches and sprinting through the station concourse (this end of work phenomenom was peculiar to the North-East and probably originated in the shipyards where men used to push against the work gates in an attempt to get away from work as soon as possible!). Occasionally a handful of miners were left on the train deeply engrossed in a card-school and the train crews had a difficult job in getting them to leave.

At one time, Westoe Lane station had the only phone in the area and locals used it frequently to phone the doctor at night, the doctor would then meet them at the station and administer to them there. It was also not unusual for the doctor to put prescriptions on the train for patients at Marsden village.

If public patronage was dwindling on the railway then coal traffic certainly was not with production at Whitburn exceeding 600,000 tons per year. In fact Whitburn broke no less than a world record for deep mine production in 1931 when 1,600 men produced 18,000 tons of coal in just one week which equated to nine loaded coal trains a day up to Westoe. Then in 1933 a dry-cleaning plant was constructed at Westoe Colliery adjacent to the electric depot. This plant separated the spoil or stone waste from the finer grades of coal and such coal

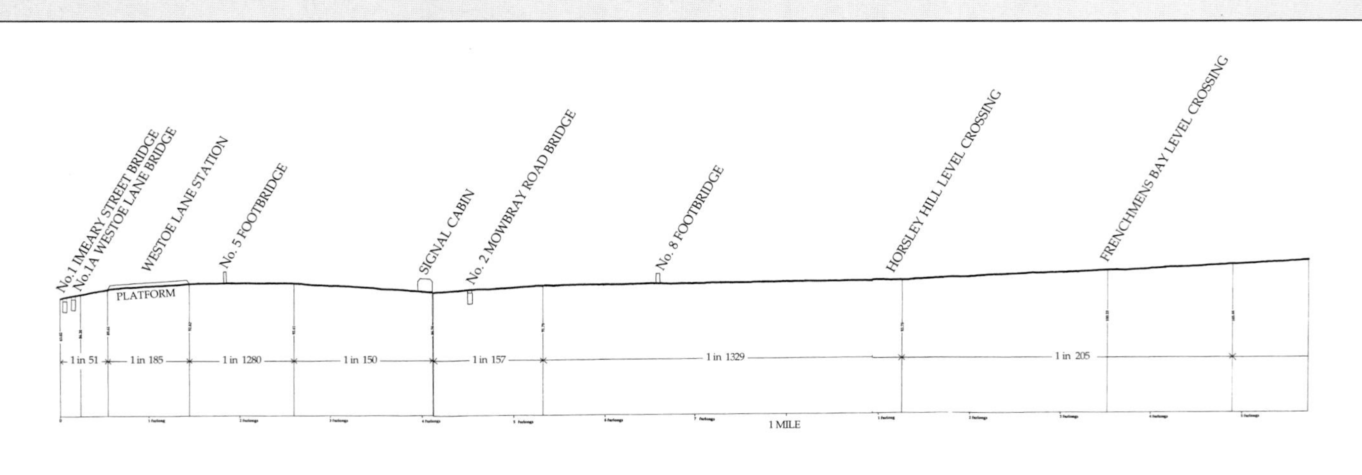

Gradient Profile

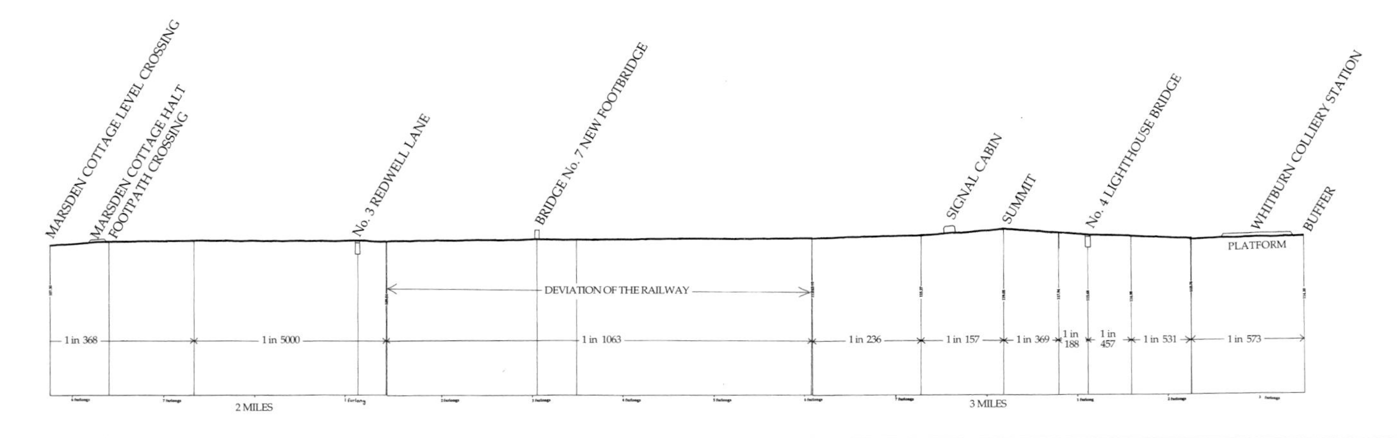

was brought from Whitburn to be shunted into the plant by electric locomotive. The plant required extensive sidings and Mowbray Road Bridge signal box had to be demolished to make way for these. This box was therefore replaced by a new one (known simply as Mowbray Road) which was situated further down the Marsden Railway between Mowbray Road bridge itself and the Trow Rocks footbridge.

A holiday camp was opened in 1936 close to Redwell Lane to cater for the vast numbers of holiday makers who were at this time flocking to Marsden Bay every summer. However any prospect of the 'Rattler' being used to cash in on this seasonal invasion was finally dashed in the same year when the Corporation introduced trolleybuses to South Shields. On 2nd May, 1937 a trolley route was extended down Redwell Lane to Marsden Bay which necessitated the replacement of the original Redwell Lane bridge with a new reinforced concrete version. Even though Redwell Lane was lowered to pass beneath the bridge, the Corporation could only utilise Metropolitan Vickers powered Karrier trolleybuses with a reduced height of 15 feet on the route. On 23rd July, 1938 a further extension saw trolleybuses running down the coast road to Marsden Bay, with the wires extending beyond this on 22nd July, 1939 to a turning circle situated somewhat ironically close to the site of Marsden station. During the summer several of these trolleybuses could be seen 'laid-up' at the Grotto awaiting passengers, their royal blue and primrose liveries adding a splash of colour to the drab coastline. The Ohio Brass Company who supplied the overhead wires described the area as 'serving the popular beach section of the community but seaswept and misty all year, and cold and blizzardly in winter'. A statement which was to portend the eventual closure of the trolley system many years later.

The Harton Coal Co. took little heed of the trolleybus invasion, and was in fact busy revamping the Lighthouse quarries. Limestone sales to the iron and steel industry had been in decline in recent years, and so the HCC was looking to produce crushed lime instead for the farming, mining, building, and chemical industries.

In August 1937, a stone crushing, screening, and storing plant was built alongside the Marsden Railway at Lighthouse bridge and was served by seven sidings. In the same year, the HCC also abandoned horse working at the quarries and replaced the 33 horses with what would eventually amount to five 2 ft gauge 4-wheel Ruston and Hornsby diesel-mechanical locomotives. One hundred men were employed at the quarries and they produced 2,000 tons of stone and 400-500 tons of lime every week.

Once the stone had been blasted from the 90 ft high working faces, steam excavators with 2 ton capacity shovels loaded the narrow gauge tubs which were then hauled to the standard gauge rail heads inside both quarries. The quarry engine then hauled the standard gauge wagons out of the quarries and over the weighbridge which was situated on the Marsden Railway directly opposite Lighthouse signal box. Any stone for crushing was then shunted into the crushing plant. The stone was hopper-discharged into the plant via a high-level siding and if any rock refused to be fed into the crusher, then it was manually sent on its way by a man levering it with an iron bar. To do this the man had to stand on the rock

The deviated railway and the new coast road

Reproduced from the 25", 1935 Ordnance Survey Map

itself, directly above the crusher. He had to watch the rock intently for tell-tale hair line cracks to appear, such cracks warning him that the rock was about to feed into the crusher and that he could easily join it if he did not jump clear!

The crushed limestone (which resembled sand) was fed into separate wagons via a low-level siding running beneath the plant. Much of this dust was fed into LNER or private owner wagons and there was a toll office at Westoe Lane station which recorded these wagon loads while they were being shunted into the 'wallside' (the track furthest from the station platform). The plant at Marsden also produced waterproof dust for the local pits and here the dust was mixed with grease which, when pumped into disused and flooded mine galleries, floated on the surface of the water and prevented the seepage of chokedamp or firedamp from the water itself. At this time lime was still produced at the kilns alongside this new outlet.

World War II had more of an effect on the Marsden Railway than the World War I ever did, although coal production at Whitburn was kept to a reasonable level at 2,733,688 tons for the period 1939 to 1945. In 1940, the HCC was forced to close down both St Hilda and Westoe collieries as an economy measure. The closure of St Hilda was permanent as the pit was exhausted, while the temporary closure at Westoe was due to the pit not being ready for full-scale coal production. The High Staiths shipping facility on the River Tyne was also closed for the duration as a further economy measure with all seabound minerals switched to the much larger Low Staiths. One thousand men were laid off due to these closures and some of these found further employment at Whitburn.

On 20th August, 1941 the Ministry of War sealed off the seafront and closed the coast road as a defensive measure with all trolleybus services along this route suspended for the duration. Heroically the 'Marsden Rattler' continued to provide a full passenger service at this time and in fact provided more trains than it had ever done before!

Passenger Timetable for 1941

Weekday trains depart Marsden
12.10, 2.30, 4.10, 4.55, 5.53, 7.15, 9.40 am, 12.00 noon, 1.10, 2.40, 3.10, 3.50, 6.35, 8.25, 9.05, 10.55 pm (The 2.30 am did not run on Mondays.)

Weekday trains depart South Shields
12.30, 3.55, 4.30, 5.10, 6.10, 9.25, 11.45 am, 12.25, 1.25, 2.55, 3.28, 5.55, 7.15, 8.40, 9.50, 10.30 pm

Saturday trains depart Marsden
2.30, 4.00, 5.10, 7.15, 8.40 am, 12.00 noon, 12.32, 1.00 pm

Saturday trains depart South Shields
3.15, 4.30, 5.30, 7.32, 8.55 am, 12.15, 12.45, 1.20 pm

Sunday trains depart Marsden
4.45, 5.20, 6.15, 9.45 am, 12.30, 1.00, 9.00, 9.50, 10.40 pm

Sunday trains depart South Shields
5.05, 5.35, 6.40, 10.15 am, 12.45, 1.20, 9.30, 10.20 pm

An unidentified 'Stubby Hawthorn' brings a 'Rattler' service into Westoe Lane station on 28th December, 1952 with miners already detraining and racing across the platform. Here it can clearly be seen how far the platform width was extended in 1900. *M.N. Bland*

Robert Stephenson & Hawthorn No. 7603 stands at Westoe Lane with the 'Rattler' on 7th August, 1952. *Alan Snowden*

As the threat of air raids became ever more real, the blackout was put into force as a counter measure and all locomotives on the Marsden Railway were equipped with anti-light emitting firebox door guards. Despite this, the 8.25 pm from Marsden was chased one night by a dogged, though fortunately inaccurate member of the Luftwaffe. Bursts of maching-gun fire accompanied the 'Rattler' along the exposed stretches of the railway with the aircraft finally disengaging as the train scuttled undamaged into Westoe Lane station! In fact the Marsden Railway never suffered any serious damage during the entire war, although the market place in South Shields was badly damaged in an October air raid of some ferocity in 1941 when three trolleybuses were destroyed. Westoe Lane station was earmarked as a collection point for the dead during the air raids and the railway saw grim times indeed during this period.

By 1944 the threat from air raids had diminished enough for the Ministry of War to re-open the coast road again and this it did on 24th June. At this time there were many American service men stationed in the area and several US rail enthusiasts visited the branch line to buy tickets as souveniers.

There was further unrest at Whitburn Colliery in 1944 with the Harton Coal Co. threatening to close the entire pit if the miners did not produce more coal, work longer hours, and mix less stone waste with the coal. By 1945, the situation was so bad at the pit that the HCC shut down the main Hutton seam putting 262 men out of work, citing strikes and unsatisfactory outputs as the cause. The Hutton seam remained unworked for three months.

After ruling the Harton empire for half a century, the Harton Coal Co. disappeared for ever at midnight on 31st December, 1946. For the Harton pits, in South Shields, however, the nationalisation of the coal industry brought little immediate change, except that is for the Marsden Railway. This stood alone in 1947 as the only nationalised public passenger railway in the country. In typical corporate style, the National Coal Board (NCB) gave the Harton system the new and rather faceless title 'No. 1 Area - North East Durham'.

Westoe Colliery was re-opened in 1947 as the NCB looked to capitalise on the vast off-shore reserves of coal so jealously guarded by the former Harton Coal Co. The letters NCB began to appear everywhere, on locomotives, carriages, wagons, uniforms, and eventually train tickets though, in the interim, station staff were instructed to cross out the name of the previous employer and write 'NCB' in ink on each ticket sold. From 1947 onwards, the NCB also began to deliver brand-new steam locomotives to the Marsden Railway, the first such locomotives to appear on the line since 1898.

The High Staiths were opened again in 1948 and were earmarked to receive all the spoil or stone waste from the No. 1 area pits for disposal at sea, thus leaving the Low Staiths free to receive train loads of coal.

The Whitburn miners must have been mightily relieved to see the arrival of underground locomotives in 1948. Prior to this, they travelled the 3¼ miles by train from South Shields only to be faced with an incredible 7 mile walk from the shaft bottom at Whitburn to the working faces out under the North Sea! The engines concerned were 2 ft 6 in. gauge 100 hp flame proof North British 0-4-0 diesel-mechanical locomotives which became something of a rarity with regards to underground engine designs.

A post-nationalisation view of Marsden Cottage Halt. Note the trolleybus overhead in the background.
Alan Snowden

This photograph is interesting in that it shows a close-up view of the original HCC shelter flanked by later NCB sheltering. Marsden limestone appears to have been used in the platform construction.
William J. Skillern

By 1950, the NCB was looking to reorganise its surface locomotives and did this by setting up a centralised locomotive works at Lambton near Durham, supplementing this with a system of central workshops sited in each area. There was a vast amount of NCB land available to the south of Whitburn Colliery so the No. 1 Area central workshops were set up here on the site of the former paper mill which had closed in 1934. In fact some of the surviving mill buildings were incorporated into the new workshops.

Marsden locomotives could now be repaired at any one of three facilities. Light and running repairs were undertaken at Whitburn Colliery shed with heavier repairs (such as serious boiler repairs) handed over to the No. 1 Area Central Workshops, while wholesale overhauls and locomotive exchanges normally involved a trip to Lambton. The No. 1 Area central workshops also constructed such items as 3 ft steel props for underground use (which in fact were made in vast quantities). The workshops were later extended to provide a facility known as withdrawn machinery stores which received such items as redundant underground locomotives.

After the war, South Shields began to see some shifts in population with miners moving from their former terraced housing close to Westoe and St Hilda collieries out to more spacious semi-detached dwellings at estates such as Horsley Hill. By 1950 so many of these miners were using Marsden Cottage Halt to catch the train to Whitburn Colliery that the Halt had to be improved. The former brick-built shelter was thus flanked by two new shelters of corrugated iron construction which ran to each end of the short platform.

In 1953 the NCB announced plans to re-build and up-grade the colliery at Westoe as part of an ambitious plan to tap into the North Sea reserves. The first stage of this plan was the construction of a vast £1 million coal preparation plant (or washer as it became known) designed to receive the new 21 ton steel-bodied wagons which were at the time being constructed in their thousands. The washery was designed to handle 500 tons of coal every hour and could easily cope with not only the combined production of the entire No. 1 Area, but also the production from pits outside the Harton network. This 'foreign' coal as it became known would be delivered to Dean Road sidings via British Railways, and brought up Chichester Road Bank and through Westoe Lane station by new 400 hp NCB electric locomotives.

For the bottleneck that was Westoe Lane station, this meant a huge increase in train movements, and so to provide an unobstructed path for the electric trains, it was decided to withdraw the South Shields, Marsden & Whitburn Colliery Railway passenger service.

The 'Rattler' soldiered on through its final year in time honoured fashion with its third class tickets still for sale at the 1888 prices of 4*d*. single and 6*d*. return! Passengers however had to seek local announcements if they were to travel on a Saturday or Sunday, such was the sporadic and haphazard nature of the services at a weekend. The final timetable read as follows:

Robert Stephenson & Hawthorn No. 7603 is seen at Whitburn Colliery station with the 'Rattler' on 7th August, 1952. *Alan Snowden*

A rake of ex-Great North of Scotland Railway six-wheeled stock stands rather forlornly at the weed infested platform at Whitburn Colliery in 1953. Note the station nameboard, high level pit sidings, and pithead bath footbridge above the carriages. *Ken Plant*

The quarry engine No. 10 (Robert Stephenson & Hawthorn 7339 of 1947) stands outside the Whitburn shed on 20th June, 1954 while one of the shed staff indulges in some ash removal. Formerly of arched construction, the shed portals were later rebuilt to allow more air to circulate through the smokey interior. *J.P.R. Bennett*

A page from the Harton Coal Co. Westoe Lane station ticket ledger.

WESTOE LANE STATION

THE HARTON COAL COMPANY, LIMITED.

SOUTH SHIELDS, MARSDEN AND WHITBURN COLLIERY RAILWAY

An Account of Tickets Issued at Westoe Lane Station on .... *day, the* .... *day of* 19

	TO MARSDEN STATION													
	RETURN							SINGLE						
	Tickets Issued				AMOUNT			Tickets Issued				Amount		
	From	To	Total	At	£	s.	d.	From	To	Total	At	£	s.	d.
First Class — Adult														
Do. Child														
Third Class— Adult														
Do. Child														
To Marsden Cottage—Adult														
Do. Child														
From Marsden Cott.—Adult														
Do. Child														
Workmen's Weekly Passes														
School Children's Tickets														
Special Meeting Tickets														
Dog Tickets														
Prams and Bicycles														
Total this day														
Do. formerly														
Do. this month														
Do. previous months														
Do. this year														

	PARCELS ACCOUNT.					LEFT LUGGAGE.						Passengers Without Tickets			Grand Total		
	NAME	Weight	Amount			Tickets Issued			Amount								
		Lbs.	£	s.	d.	From	To	Total	£	s.	d.	£	s.	d.	£	s.	d.
First Class — Adult																	
Do. Child																	
Third Class— Adult																	
Do. Child																	
To Marsden Cottage—Adult																	
Do. Child																	
From Marsden Cott.—Adult																	
Do. Child																	
Workmen's Weekly Passes																	
School Children's Tickets																	
Special Meeting Tickets																	
Dog Tickets																	
Prams and Bicycles																	
Total this day																	
Do. formerly																	
Do. this month																	
Do. previous months																	
Do. this year																	

Total Banked This Day

Do. Formerly

Do. This Month

.... Station Master

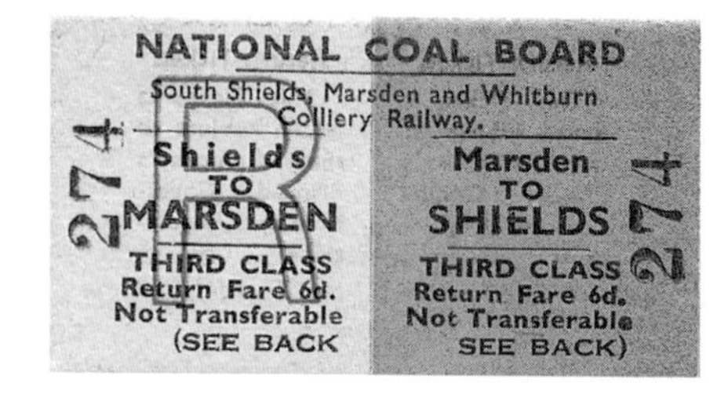

Passenger Timetable for 1953

Weekday trains depart Marsden
12.30, 2.00, 4.55, 6.00, 7.10, 10.40 am, 1.35, 2.15, 3.05, 4.48, 6.35, 9.20 pm

Weekday trains depart South Shields
1.45, 4.25, 5.43, 6.25, 9.28 am, 12.05, 2.00, 2.48, 3.27, 5.15, 8.03, 9.58 pm

Many die-hard enthusiasts travelled up to South Shields in this final year for one last journey on the 'Rattler' and found Westoe Lane station to be a rather melancholy place bereft of name boards and indeed any other public notices.

The service was replaced by a fleet of buses which began taking miners down to Whitburn Colliery on Monday 14th November, 1953. But for some strange reason the 'Rattler' continued to operate for the general public for just one more week beyond this while the NCB announced that there would be a special commemorative train later in the year (a promise in fact, which never came to fruition).

Sunday 22d November, 1953 saw the final 'Rattler' service waiting to depart from a deserted Westoe Lane station at 9.58 pm. With its miners now travelling to work by bus, and enthusiasts staying away and awaiting a last commemorative train which was not to be, the sad fact of the matter is, that as the train disappeared round the curve beyond Westoe and into the history books, it was probably empty.

PUBLIC NOTICE

South Shields, Marsden & Whitburn Colliery Light Railway

As alternative, frequent road services are available to the public between South Shields and Whitburn Colliery the National Coal Board have decided to discontinue the passenger train service on this railway on and from Monday, 23rd November, 1953.

National Coal Board,
Durham Division,
No 1 Area
Station Road,
South Shields.

Newspaper cutting from the 13th November, 1953 edition of the *Shields Gazette*.

Shields Gazette

Chapter Four

Locomotives, Carriages and Wagons

The very term 'Light Railway' conjures up visions of a quiet, half forgotten backwater of a line served over the decades perhaps by a single wheezing tank engine plying its way to and fro with its faithful little rake of carriages in tow. But not so the Marsden Railway.

In its 89 years of operations, the 3¼ miles of track running between South Shields and Marsden was host to an almost bewildering variety of rolling stock with a new locomotive appearing on the railway, on average, every 2 years!

During the initial 12 years (1879-1891) when the Whitburn Coal Company were made responsible for 'kick-starting' the railway into life, five 0-6-0 saddle tanks were employed. Official WCC livery is not known but what is known is that lining was used and that lettering was kept to a minimum, with the number only of each locomotive appearing on the tank sides.

SSMWCR Locomotives introduced by the Whitburn Coal Company

Engine No.:	1	*Cylinders:*	12 in. x 18 in. inside
Builder:	Manning, Wardle	*Weight:*	
Works No.:		*Driving wheels:*	3 ft 0 in.
Built:		*Boiler pressure:*	
Type:	0-6-0ST	*Fate:*	Scrapped 1895-96

The origins of this locomotive (the first to work on the Marsden Railway) are lost in time, however, it is known that is was rebuilt by the Grange Iron Co. Ltd at its Belmont Works near Durham in 1875.

The Grange Iron Co. specialised in structural, as well as general engineering and in fact was employed to supply the main sections for the road bridges along the railway. It is not inconceivable, therefore, that the locomotive may have been employed on the Marsden Railway initially by the Grange Iron Co. when the bridge sections were being laid in place, possibly being sold on to the WCC when the work was completed.

Engine No.:	2	*Cylinders:*	14 in. x 20 in. outside
Builder:	Black, Hawthorn	*Weight:*	
Works No.:	504	*Driving wheels:*	3 ft 6 in.
Built:	1879	*Boiler pressure:*	
Type:	0-6-0ST	*Fate:*	Scrapped 1905

Purchased new by the Whitburn Coal Co., an order was placed at the builders for this locomotive on 11th June, 1879. No. 2 went on to haul the first 'Rattler' services.

All the dimensions and details listed in this chapter pertain to each locomotive as it was introduced onto the Marsden Railway and in many cases this differs from locomotive dimensions as built. The weights listed are full working weights (i.e. with fully coaled and watered, tenders etc.).

The Whitburn Coal Company purchased just five locomotives for the Marsden Railway. No. 4 was the most succesfull of the five becoming the only Marsden locomotive to be utilised by the Whitburn Coal Co., the Harton Coal Co., and the NCB. Here she is seen back in 1884 in ex-works condition. Note the extensive use of double lining, but lack of company lettering.

Author's Collection

The ex-Whitburn Coal Co. engine No. 4 (Black, Hawthorn 826 of 1884) obscures the water column as she stands outside Whitburn shed on 1st April, 1934. The platelayer's trolleys had to be derailed between duties in order to prevent any potential runaways. *H.C. Casserley*

Engine No.:	3	*Cylinders:*	15 in. x 22 in. inside
Builder:	Black, Hawthorn	*Weight:*	32 tons
Works No.:	716	*Driving wheels:*	3 ft 9½ in.
Built:	1882	*Boiler pressure:*	
Type:	0-6-0ST	*Fate:*	Scrapped, date unknown

Black, Hawthorn specifically designed this 12 feet wheelbase locomotive for what they termed 'higher speed work on an easy curved branch'. It was designed to haul 729 ton trains on level track at 12-15 mph or 76 ton trains on a gradient of 1 in 25 at the same speed. It survived into Harton Coal Co. ownership and was rebuilt with 14 in. x 20 in. cylinders and to 'cut down' format in 1906 before being transferred down to St Hilda Colliery where it worked the ½ mile Low Staiths branch (a section of track which ran through a low, narrow tunnel). This branch was electrified in 1908 and so No. 3 was sold in 1910 to James W. Ellis & Co.'s Foundry at Swalwell. The locomotive subsequently returned to pit work being employed at Usworth Colliery and later still Ryhope Colliery.

Engine No.:	4	*Cylinders:*	15 in. x 22 in. inside
Builder:	Black, Hawthorn	*Weight:*	32 tons
Works No.:	826	*Driving wheels:*	3 ft 9½ in.
Built:	1884	*Boiler pressure:*	
Type:	0-6-0ST	*Fate:*	Scrapped at Whitburn Colliery in 1948

An identical locomotive to No. 3, No. 4 was ordered from Black, Hawthorn on 27th May, 1884 with delivery promised within 3½ months. An interesting design concept involving both No. 3 and No. 4 was that there appeared to be only one entrance to the cab (presumably to give the crews some protection from the North Sea weather). No. 4 also survived into Harton Coal Co. ownership and was bestowed with an SSMWCR number plate in place of her painted number. These number plates featured on many of the Marsden Railway locomotives and consisted of the running number in the centre of the plate, while bordering the number in smaller lettering were the words: 'THE SOUTH SHIELDS MARSDEN AND WHITBURN COLLIERY RAILWAY'. No. 4 was also rebuilt to 'cut down format' in 1906 and re-allocated alongside No. 3. However, No. 4 remained in this form until 1923 (and was presumably kept as a spare locomotive in case of problems with the overhead wires in the Low Staiths tunnel). The locomotive was then rebuilt again in 1923 back to her original state apart from the addition of a steam dome.

The engine was then used as a spare quarry locomotive and just survived into National Coal Board ownership. With 64 years' service under her belt, she in fact became the longest working of the Marsden locomotives.

Engine No.:	5	*Cylinders:*	15 in. x 22 in. inside
Builder:	Robert Stephenson	*Weight:*	
Works No.:	2629	*Driving wheels:*	
Built:	1887	*Boiler pressure:*	
Type:	0-6-0ST	*Fate:*	Scrapped 1922

A little-known locomotive purchased new for the Whitburn Coal Co., No. 5 was employed at the Marsden Quarries until 1922.

SSMWCR Locomotives introduced by the Harton Coal Company

The Harton Coal Co. took full control of the Marsden Railway in 1891 and the long period of austerity instigated by this rather frugal company began. After buying its initial Marsden locomotive brand new, the HCC reverted to a policy of second-hand only. Much of this used stock was in the form of ex-North Eastern Railway 0-6-0 tender engines of various classes, which became known on the railway as 'the big engines'. Although the brass number plates were seen on some of the locomotives, many simply ran with painted numbers and lettering while the livery was unlined black. The Harton Coal Co. made many of its own spare parts for the locomotives and, before an engine was scrapped, it was stripped of all reusable parts (even nuts and bolts were polished up and stored).

If the train crews were lucky, the ex-NER locomotives arrived with tender cabs. There were no turntables on the railway and locomotives were normally run tender first down to Whitburn and a working day spent thus in an open-backed cab in mid-winter must have been unpleasant indeed.

The first HCC locomotive purchased for the Marsden Railway was numbered 7. Quite why there was a gap in the numbering from the previous Whitburn Coal Co. engines is inexplicable. If there indeed was a No. 6, then no detail of it survives.

When the HCC withdrew a locomotive, its number was automatically transferred onto the next incoming engine. So even though the HCC employed 23 steam locomotives on the Marsden Railway during its 56 year reign, numbering never went above number 11.

Engine No.:	7	*Cylinders:*	17 in. x 24 in. inside
Builder:	Chapman & Furneaux	*Weight:*	
Works No.:	1158	*Driving wheels:*	4 ft 3 in.
Built:	1898	*Boiler pressure:*	
Type:	0-6-2T	*Fate:*	Scrapped on the Pontop & Jarrow Railway at Springwell 1923

With her 0-6-2 wheel arrangement, tall steam dome and chimney, twin cab-side windows, and masses of fine lining art work, No. 7 was something of an elegant departure in design terms from the rugged 0-6-0 saddle tanks then extant on the railway. The locomotive was ordered from Chapman & Furneaux (the successors to Black, Hawthorn) on March 1898 with delivery promised

No. 7 was the only locomotive to be purchased new by the Harton Coal Co. Curiously elegant by colliery standards, with much fine lining and SSMWCR number plate in evidence, this Chapman & Furneaux design of 1898 led a short and troubled existence until she was scrapped in 1923. Note also the double features of drawbars, buffers, and cabside windows. *R.M. Casserley Collection*

within 6 months. No. 7 also carried several unusual additions such as twin drawbars (one above the other) and both spring and dumb buffers designed to be compatible with both the 10½ ton wagons and the dumb-buffered chaldron wagons. The engine was equipped with a steam brake while the pony wheels were supported by a radial truck.

Despite her elegant appearance, internally No. 7 suffered. On Boxing Day 1906, she disgraced herself by blowing a tube and badly injuring her crew. After subsequent problems she was sold in 1912, to Robert Frazer & Sons Ltd (an Hebburn-based company which dealt in rolling stock). Robert Frazer sold the locomotive on to John Bowes & Partners' Pontop & Jarrow Railway where it joined three remarkably similar stablemates. The locomotive was renumbered No. 14 and continued to suffer a series of faults until withdrawn.

Engine No.:	8	*Cylinders:*	16 in. x 28 in. inside
Builder:	Sharp, Stewart	*Weight:*	30 tons 5 cwt
Works No.:	1501	*Driving wheels:*	5 ft 6 in.
Built:	1864	*Boiler pressure:*	120 psi
Type:	2-2-2WT	*Fate:*	Scrapped 1907

Originally a Furness Railway 'B3' class engine built at the Manchester works of Sharp, Stewart, No. 8 began life as No. 22 (the number being carried on a brass plate). This was another elegant locomotive with a huge brass steam dome, stove pipe chimney, and large single driving wheels, all set off with a livery of black and Indian red. The Furness Railway sold the locomotive as No. 22A to Robert Frazer in August 1899 who then sold it on to the Harton Coal Co. in the same year. Apparently the locomotive found little favour with the Marsden train crews and it was withdrawn eight years later.

Engine No.:	9	*Cylinders:*	16 in. x 24 in. inside
Builder:	Blyth & Tyne Railway	*Weight:*	
Works No.:		*Driving wheels:*	4 ft 6 in.
Built:	1862	*Boiler pressure:*	140 psi
Type:	0-6-0	*Fate:*	Scrapped 1913

This locomotive was one of just a handful built at the Percy Main works of the enigmatic Blyth & Tyne Railway. Water was in limited supply along the Blyth & Tyne so rather than build tank engines, the company equipped these designs with 1,500 gallon capacity, 6-wheel tenders. Although the Blyth & Tyne was later absorbed by the NER, this locomotive was immediately distinguishable from the NER 0-6-0 designs by its curiously angular cab and sloping cab roof. Originally numbered No. 3 by the Blyth & Tyne, it was renumbered no less than four times by the NER (No. 1303 in August 1874, No. 1923 in August 1891, No. 1733 in January 1894, and finally No. 2255 in March 1899). It was withdrawn in April 1900 by the NER and sold directly to the HCC in May of the same year where it became No. 9.

This was the first in a long line of NER 0-6-0 tender engines to work on the HCC and in the ensuing years the Marsden Railway came to display quite a little showcase of North Eastern design advances in this type. No. 9 was also the first locomotive to be purchased by the HCC specifically as a bulk mineral engine, its delivery coinciding with that of the first NER pattern 10½ ton spring-buffered wagons which were to become so much a feature of the Marsden Railway.

Engine No.:	10	*Cylinders:*	16 in. x 24 in. inside
Builder:	Blyth & Tyne Railway	*Weight:*	
Works No.:		*Driving wheels:*	4 ft 6 in.
Built:	1862	*Boiler pressure:*	140 psi
Type:	0-6-0	*Fate:*	Scrapped 1914

The HCC must have been impressed with No. 9, because in November 1900 they purchased a second member of the ex-NER 'Blyth & Tyne' class. This locomotive had a similar history to that of No. 9, beginning life, in this case as No. 14, before enduring several renumberings under the NER (namely No. 1314 in August 1874, No. 1729 in July 1892, and No. 1712 in January 1894).

While enjoying an extended existence on the Marsden Railway, No. 10 had the distinction of becoming the oldest working North Eastern Railway locomotive then in existence.

Engine No.:	6	*Cylinders:*	17 in. x 24 in. inside
Builder:	Robert Stephenson	*Weight:*	66 tons 4 cwt
Works No.:	2160	*Driving wheels:*	5 ft 0 in.
Built:	1874	*Boiler pressure:*	140 psi
Type:	0-6-0	*Fate:*	Scrapped 1912

No. 6 was a very early member of the 325 strong NER '398' class of locomotives and was in fact part of a second batch of 30 locomotives built under contract by Robert Stephenson at its Forth Street Works in Newcastle in February 1874. She was numbered No. 888 by the NER and carried this number until withdrawn. The engine was then purchased by Robert Frazer on 18th December, 1907 who immediately sold it on to the HCC. No. 888 thus escaped the complex rebuilding which affected later members of the class and arrived on the Marsden Railway complete with original 'telescopic' boiler and longer 12 ft 3 in. wheel base tender. The cab was also an early design with round, hooded spectacles and sweeping cut away cab sides. No. 6 had slotted frames and Worsdell cast-iron chimney.

In common with the Blyth & Tyne engines, No. 6 was given bulk mineral duties on the Marsden Railway.

Engine No.:	8	*Cylinders:*	17 in. x 24 in. inside
Builder:	Robert Stephenson	*Weight:*	66 tons 4 cwt
Works No.:	1973	*Driving wheels:*	5 ft 0 in.
Built:	1870	*Boiler pressure:*	140 psi
Type:	0-6-0	*Fate:*	Scrapped 1929

The '708' class were the last in a line of NER double-framed locomotives and were represented on the Marsden Railway by just one example, No. 718. Built under contract by Robert Stephenson's, No. 718 featured a tender complete with cab and four coal rails. The NER also equipped the

No. 8 was one of the grand old ladies of Marsden being built by Robert Stephenson and Co. back in 1870. Thus, by the time this August 1924 photograph was taken at Whitburn Colliery, this ex-NER '708' class locomotive had already put in over 50 years' service. The Westinghouse Brake equipment which allowed No. 8 to haul the 'Rattler' can be seen just forward of the cab. *R.M. Casserley Collection*

locomotive with a Westinghouse air brake sometime in the 1880s and when it was sold to Robert Frazer on 18th December, 1907, Frazer immediately sold it on to the HCC as a mixed traffic design capable of both heavy coal and 'Rattler' duties on the Marsden Railway where it ran complete with brass number plate.

Engine No.:	11	*Cylinders:*	12 in. x 18 in. outside
Builder:	Manning, Wardle	*Weight:*	19 tons 10 cwt
Works No.:		*Driving wheels:*	3 ft 6 in.
Built:		*Boiler pressure:*	
Type:	0-4-0ST	*Fate:*	Scrapped 1920

As can be seen, little is known about this curiously small (by HCC standards) 0-4-0 design. It was purchased second-hand by the HCC from an unknown source in 1908, possibly for use as a colliery shunter marshalling trains for the big engines. No. 11 was the locomotive which hauled the armoured train along the Marsden Railway during World War I.

Engine No.:	6	*Cylinders:*	17 in. x 24 in. inside
Builder:	Robert Stephenson	*Weight:*	66 tons 5 cwt
Works No.:	2056	*Driving wheels:*	5 ft 0 in.
Built:	1872	*Boiler pressure:*	140 psi
Type:	0-6-0	*Fate:*	Scrapped January 1930

Another early NER '398' class locomotive built under contract in June 1872, No. 6 began life as No. 786. As with the previous No. 6, it was largely unrebuilt and featured a telescopic boiler, 12 ft 3 in. wheelbase tender, slotted frames, and a Worsdell cast-iron chimney. No. 786 was sold to Robert Frazer on 27th March, 1912 who duly sold it on to the HCC in June 1912. No. 6 was withdrawn in 1927.

Engine No.:	10	*Cylinders:*	17 in. x 24 in. inside
Builder:	R. & W. Hawthorn	*Weight:*	66 tons 4 cwt
Works No.:	1564	*Driving wheels:*	5 ft 0 in.
Built:	1873	*Boiler pressure:*	140 psi
Type:	0-6-0	*Fate:*	Scrapped 1931

One of a later batch of just 20 class '398s' built under contract by R. & W. Hawthorn, No. 10 was originally numbered No. 827 by the NER. She was rebuilt at York Works in 1886 with a 'Fletcher' type iron boiler. The HCC purchased the locomotive from the NER on 31st August, 1914 and employed her on the Marsden Railway until 1931.

This Gateshead built ex-NER '398' class engine ran on the Marsden Railway as No. 6 between 1927 and 1930. However, here it is seen in later guise as No. 1 at Boldon Colliery in the 1930s. These rebuilt engines always looked rather uncomfortable when re-coupled to their unrebuilt tenders. *Frank Jones*

Engine No.:	6	*Cylinders:*	17½ in. x 24 in. inside
Builder:	NER	*Weight:*	67 tons 18 cwt
Works No.:		*Driving wheels:*	5 ft 0 in.
Built:	1882	*Boiler pressure:*	160 psi
Type:	0-6-0	*Fate:*	Scrapped *c.*1940

This was a much later class '398' locomotive built by the NER at Gateshead works in November 1882. Coincidentally the locomotive originally ran on the NER as No. 6 before being renumbered No. 1453 in 1914. No. 6 was rebuilt with a Worsdell steel boiler in January 1898 and fitted with 17 ½ inch bore cylinders in June 1923.

The LNER sold No. 1453 to the HCC in February 1927 where it ran complete with steam brake, double side windows to the cab, and a sandwich-framed tender with two coal rails. In June 1930, the HCC transferred No. 6 to Boldon Colliery from where it worked coal trains up to the Harton Railway by means of a running powers agreement with the LNER.

Engine No.:	5	*Cylinders:*	17 in. x 24 in. inside
Builder:	NER	*Weight:*	66 tons 14 cwt
Works No.:		*Driving wheels:*	5 ft 1 in.
Built:	1881	*Boiler pressure:*	160 psi
Type:	0-6-0	*Fate:*	Scrapped at Whitburn by contractors on 28th February, 1953

No. 5 began life as No. 396 and was another Gateshead '398' class construction. It was rebuilt with a Worsdell steel boiler in February 1900 and was also fitted at some time with both steam and Westinghouse brakes, and a tender with steel plates rather than coal rails.

The ex-NER '398' class locomotive of 1881 No. 5 is seen between duties at Whitburn shed. The redundant 'Lancashire' and 'Cornish' boilers seen positioned at the end of the shed building provided water for the many duty locomotives allocated to the pit. This photograph was taken on 29th April, 1952 less than a year before No. 5 was scrapped. *H.C. Casserley*

The '398' class engine No. 5 is just one of several locomotives receiving attention inside a rather cluttered looking Whitburn shed on 9th June, 1934. The footplate detail is interesting. *H.C. Casserley*

The engine was sold by the NER to Frazer & Son in August 1925 for the princely sum of £800. The HCC in turn purchased the locomotive for the SSMWCR in October 1929 where it initially ran as No. 11 before being renumbered No. 5 (this being carried on a brass plate).

No. 5 had what was described as 'a tight boiler' by her driver but had no tender cab and so was confined to colliery shunting at Whitburn, being used on the 'Rattler' only if No. 8 or No. 6 were unavailable. No. 5 survived Nationalisation to become the last working '398' class locomotive in existence.

Engine No.:	7	*Cylinders:*	18 in. x 24 in. inside
Builder:	NER	*Weight:*	77 tons 6 cwt
Works No.:	38 (1892)	*Driving wheels:*	5 ft 1 in.
Built:	1892	*Boiler pressure:*	160 psi
Type:	0-6-0	*Fate:*	Scrapped 1935

This was the first of the four highly popular 'C' class (later 'J21' class under the LNER) locomotives to appear on the Marsden Railway. As No. 1616 it began life as a Gateshead built compound engine before being rebuilt as a simple engine by the NER in May 1905.

Sold directly to the HCC on October 1929, No. 7 was used on the 'Rattler' having been sold complete with Westinghouse brake.

Engine No.:	6	*Cylinders:*	17 in. x 26 in. inside
Builder:	Robert Stephenson	*Weight:*	64 tons 10 cwt
Works No.:	2587	*Driving wheels:*	5 ft 1 in.
Built:	1884	*Boiler pressure:*	160 psi
Type:	0-6-0	*Fate:*	Scrapped 1936

Another ex-NER class to appear on the railway was the class '59' (later 'J22' class under the LNER), albeit represented by just one locomotive No. 1486. This was one of the 12 engines built under contract by Robert Stephenson's, these 12 having different tender designs featuring 4 ton coal capacity and 2,500 gallon capacity tanks. No. 1486 was rebuilt in April 1900 with a Worsdell steel boiler and in January 1930 became the last of the 44 strong class to be withdrawn.

Sold direct to the HCC in January 1930, No. 6 was another Westinghouse and steam brake-fitted locomotive which worked passenger trains on the railway.

Engine No.:	8	*Cylinders:*	18 in. x 24 in. inside
Builder:	NER	*Weight:*	77 tons 6 cwt
Works No.:	3 (1889)	*Driving wheels:*	5 ft 1 in.
Built:	1889	*Boiler pressure:*	160 psi
Type:	0-6-0	*Fate:*	Scrapped at Whitburn in August 1954

No. 869 was a 'C' class engine built at Gateshead as a compound locomotive in February 1889. It was rebuilt as a simple locomotive in January 1904 and was

A drawing of No. 8 (the ex-NER 'C' class locomotive, Works No. 3 [1889]). *D. Monk-Steel*

withdrawn and sold to the HCC in January 1931, being the last of its class to be withdrawn complete with Westinghouse brake.

An extremely robust engine, No. 8 (the number was painted on the cab side) became the 'pride of the fleet' at Whitburn, her performance being summed up by her regular driver thus: 'You just had to pick up the shovel, show it to No. 8, and the steam used to fly out!' The engine put in yeoman service on the Marsden Railway and in 1951 was given a replacement boiler which originated from No. 6 (ex-NER No. 23 [1889]). No. 8 became a spare engine in April 1953.

Engine Name:	*Laleham*	*Cylinders:*	14 in. x 22 in. outside
Builder:	Andrew Barclay	*Weight:*	32 tons 6 cwt
Works No.:	1639	*Driving wheels:*	3 ft 4 in.
Built:	1922	*Boiler pressure:*	160 psi
Type:	0-6-0ST	*Fate:*	Scrapped at Boldon Colliery by J. Wright & Co. Ltd in June 1964

The construction firm S. Pearson & Son Ltd was awarded a Metropolitan Water Board Contract in 1922 to built several large reservoirs at Sunbury, Greater London. They immediately purchased Barclay 1639 (which was ex-works on 29th December, 1922) and put it on this contract. It was named *Laleham* after one of the villages in the Sunbury district (the name being carried on brass plates on the tank sides). *Laleham* was put up for sale in March 1929 and was purchased by another contractor, Geo. Cohen & Sons Ltd, who sold it to the HCC in the same year.

The erstwhile Marsden Quarry engine *Laleham* stands between duties at Whitburn shed on 1st April, 1934. Her nameplate is clearly evident as is a 'blank' numberplate on the cabside and Barclay worksplate on the side of the smokebox. Note also the rather ponderous dumb buffers. *H.C. Casserley*

Originally employed as a shunter at Boldon Colliery, *Laleham* was transferred to Whitburn in July 1931. It was a spare engine being put to work at Marsden Quarries when No. 4 (Black, Hawthorn No. 826) was unavailable. The NCB transferred *Laleham* back to Boldon in February 1960 where it was allocated as No. 1 (a number it never actually carried) and repainted black with red lining. *Laleham* became the last ex-HCC steam locomotive to work on the Harton system.

Engine No.:	7	*Cylinders:*	18 in. x 24 in. inside
Builder:	NER	*Weight:*	77 tons 6 cwt
Works No.:	43 (1889)	*Driving wheels:*	5 ft 1 in.
Built:	1889	*Boiler pressure:*	160 psi
Type:	0-6-0	*Fate:*	Scrapped 1939

No. 776 was built as a 'C' class compound locomotive at Gateshead and was rebuilt as a simple locomotive in December 1904. It was also fitted with steam, vacuum and Westinghouse brakes (the latter being fitted in June 1923). On 17th May, 1935, the locomotive was withdrawn and sold to the HCC becoming No. 7 (the number being painted on the cab side).

Engine No.:	6	*Cylinders:*	18 in. x 24 in. inside
Builder:	NER	*Weight:*	74 tons 12 cwt
Works No.:	23 (1889)	*Driving wheels:*	5 ft 1 in.
Built:	1889	*Boiler pressure:*	160 psi
Type:	0-6-0	*Fate:*	Scrapped at Whitburn Colliery in June 1951

No. 1509 was the last of the ex-NER 'C' class locomotives put to work on the Marsden Railway. Originally a compound, it was rebuilt as a simple locomotive in December 1904. No. 1509 was fitted with a class '398' boiler *circa* 1929 and was withdrawn and sold to the HCC on 2nd August, 1935.

Although it was equipped with a Westinghouse brake, No. 6 was employed mainly as a pilot engine at Whitburn Colliery being used on the 'Rattler' only if No. 8 (NER 3 [1889]) was unavailable. No. 6 was withdrawn in 1951 and her class '398' boiler lifted out and refitted to No. 8 in March of that year.

Engine No.:	7	*Cylinders:*	18½ in. x 24 in. inside
Builder:	NER	*Weight:*	73 tons 14 cwt
Works No.:	631 (1898)	*Driving wheels:*	4 ft 7 in.
Built:	1898	*Boiler pressure:*	160 psi
Type:	0-6-0	*Fate:*	Scrapped 1946

Yet another NER design to find employment on the Marsden Railway was the 'P' class (reclassified as 'J24' by the LNER), a smaller-wheeled version of the 'C' class. No. 1953 was rebuilt in March 1915 with a superheated boiler and piston valves, and was withdrawn and sold to the HCC on 20th May, 1939.

Bereft of a Westinghouse brake, No. 7 was used as a pilot engine at Whitburn Colliery.

SSMWCR Locomotives introduced by the NCB

In 1947, the NCB found itself inheriting a fleet of steam locomotives at Whitburn Colliery that were both aged and overworked in the extreme. These wheezing giants leaked steam and water from tubes, roofboards, and seams and were in dire need of replacement. This crisis which blighted the early years of Nationalisation was not fully remedied until 1954, and in the interim period locomotives had to be drafted in from other NCB systems and even hired from British Railways.

The NCB also eventually abandoned the existing livery of unlined black replacing it with its own livery of navy blue with yellow numbering. One of the miners at Whitburn was something of an artist in his spare time and was regularly employed at Whitburn shed whenever a locomotive needed repainting. To his credit, this gentleman took the care to embellish the locomotives with extensive fine lining which in some cases gave the engines a much improved appearance.

Engine No.:	9	*Cylinders:*	18 in. x 26 in. inside
Builder:	Hunslet	*Weight:*	43 tons 3 cwt
Works No.:	3191	*Driving wheels:*	4 ft 3 in.
Built:	1944	*Boiler pressure:*	170 psi
Type:	0-6-0ST	*Fate:*	Scrapped at Wearmouth Colliery in March 1969

An austerity design, Hunslet 3191 began life as a War Department locomotive being numbered No. WD75140. After being employed at Ministry of Defence Depots at Kineton, Steventon and Lockinge, the locomotive was sold to the NCB in June 1947, and found initial employment at Whitburn Colliery. No. 9 was sent to the main NCB locomotive works at Lambton in 1951 for the fitting of a new firebox before returning to Whitburn in April 1952. No. 9 was then transferred to Boldon Colliery in 1956, returning to Whitburn in 1960 for the fitting of a Giesl ejector. This ejector was fitted at the Area central workshops in February 1961 and was incorporated as part of an NCB experiment into the possibilities of the smokeless burning of small coals. In this guise, No. 9 began work at Whitburn Colliery on 21st February, 1961.

No. 9 found little favour with her crews, as the ejector (which the crews nicknamed 'The Funkenbanger') had to be constantly cleaned out for it to work efficiently, her driver stating that: 'When you opened up the regulator, half the fire used to lift off the grate and disappear up the tubes'. In 1963, the locomotive was renumbered No. 2 and was waylaid at the Area central workshops between 10th December, 1963 and 7th May, 1965 for heavy general repairs. No. 2 was eventually transferred to the Lambton engine works.

Engine No.:	10	*Cylinders:*	16 in. x 24 in. outside
Builder:	R. Stephenson, Hawthorn	*Weight:*	43 tons 6 cwt
Works No.:	7339	*Driving wheels:*	3 ft 8 in.
Built:	1947	*Boiler pressure:*	180 psi
Type:	0-6-0ST	*Fate:*	Scrapped August 1970

No. 10 was delivered new to Whitburn Colliery on 8th October, 1947 and was in fact the first new locomotive to work on the Marsden Railway since the Chapman & Furneaux locomotive No. 7 back in 1898. Any repairs were initially undertaken by the Stephenson, Hawthorn's fitters on site at Whitburn shed. No. 10 was purchased specifically for quarry shunting at Marsden, taking over these duties from No. 4 (Black, Hawthorn No. 826) and *Laleham,* but was also employed on 'Rattler' duties (despite being equipped with neither vacuum nor air train brakes) if no other locomotive was available.

On 5th July, 1961, No. 10 was sent to the Area central workshops with a condemned firebox, then after spending the following 3½ years out of traffic the locomotive was reallocated to Whitburn shed on 17th December, 1964 resplendent in NCB navy blue livery but without running number. The locomotive was transferred to the Washington 'F' pit in March 1966 after the Marsden Quarries were sold off.

Engine No.:	1513	*Cylinders:*	13 in. x 20 in. inside
Builder:	Hudswell, Clarke	*Weight:*	24 tons 15 cwt
Works No.:	1513	*Driving wheels:*	3' 3½"
Built:	1924	*Boiler pressure:*	160 psi
Type:	0-6-0ST	*Fate:*	Scrapped at Boldon Colliery on 4th December, 1959

Works No. 1513 lead quite a nomadic existence after being delivered new to Robert McAlpine on 20th March, 1924 as No. 42, being employed on road building contracts in Glasgow and Birmingham. No. 42 was then used on the construction of both Tilbury Docks and the Southern Railway Docks at Southampton.

The locomotive was then sold to another contractor, John Mowlem in 1940 where it was named *Staines,* and was employed under two armed forces contracts at Bednall Wharf, Staffordshire, and Glynrhonwy Quarry, North Wales.

The War Department acquired the engine on 20th August, 1947, numbered it No. WD70069, and sent it to the Central Ordnance Depot at Bicester. No. WD70069 was then sent to Hudswell, Clarke for a major overhaul in December 1947, who then sold it on to the NCB on New Year's Day 1948.

Whitburn Colliery thus received it second ex-War Department locomotive (which the Whitburn crews nicknamed 'Auxiliaries'), but never actually numbered the locomotive, referring to it by the number shown on its worksplate, No. 1513. In fact this method of numbering was applied to several further NCB locomotives delivered to Whitburn. No. 1513 was looked upon as somewhat underpowered for the heavy Marsden duties and had been transferred to Boldon Colliery by September 1948, returning to Whitburn shed briefly in January 1950 for the fitting of new boiler tubes and subsequent boiler test.

Engine No.:	7132	*Cylinders:*	18 in. x 26 in. inside
Builder:	R. Stephenson, Hawthorn	*Weight:*	48 tons 3 cwt
Works No.:	7132	*Driving wheels:*	4 ft 3 in.
Built:	1944	*Boiler pressure:*	170 psi
Type:	0-6-0ST	*Fate:*	Scrapped at Wearmouth Colliery in March 1969

Another War Department Austerity, this engine was delivered new to the Melbourne Military Railway as No. WD75182. It was registered with the Great Western Railway in 1945 and put to work at Treforest and Maesmawr in South Wales before being sold to the NCB for Boldon Colliery duties in July 1947.

No. 7132 finally arrived at Whitburn Colliery via a transfer on 20th September, 1948 and was later transferred back to Boldon on 6th May, 1954. No. 7132 received a new firebox in 1961 and was renumbered No. 5 in 1963. The engine was then sent to the Area central workshops at Whitburn on 28th September, 1963 for a heavy general overhaul, new boiler and firebox, and the fitting of a Hunslet underfeed stoker and gas producer system, before returning to duties at Whitburn shed on 14th June, 1964. The underfeed stoker was fitted to reduce the dark smoke which normally plagued the burning of small coals in the firebox and (as with the Giesl ejector) found little favour with the train crews at Whitburn. Frequently, the fireman used to open up the underfeed stoker and then forget to shut it down again so that later when he went to inspect the fire, he found himself confronted by a wall of fresh coal, with the fire crushed against the roof of the firebox! The engine was working with the stoker out of use by August 1965 and on Christmas Day 1965 the locomotive was transferred to Boldon Colliery. It never returned.

Engine No.:	7603	*Cylinders:*	18 in. x 24 in. outside
Builder:	R. Stephenson, Hawthorn	*Weight:*	53 tons
Works No.:	7603	*Driving wheels:*	4 ft 0 in.
Built:	1949	*Boiler pressure:*	180 psi
Type:	0-6-0ST	*Fate:*	Scrapped at Derwenthaugh shed, NCB Swalwell in October 1972

In 1949, the NCB decided to introduce a standard mixed traffic engine specifically for the Marsden Railway which would need to be not only powerful enough to cope with heavy mineral trains but also equipped to head the passenger service, thus allowing Whitburn to withdraw the life expired 'Big Engines'.

Robert Stephenson & Hawthorn was awarded the contract and took a design which had found success at the steel works of Dorman, Long & Co., and added to this design Westinghouse and vacuum brakes. All the carriages on the

Robert Stephenson & Hawthorn No. 7132 of 1944 was fitted with a Hunslet underfeed stoker and gas producer system in 1964 and henceforward became instantly recognisable on account of the redesigned chimney. This is how she looked on 6th October, 1964 shortly after emerging from the rebuild. *Industrial Railway Society/Brian Webb Collection*

Robert Stephenson & Hawthorn No. 7603 is seen taking water at Westoe Lane station on 7th August, 1952. *Alan Snowden*

Marsden Railway were in fact air braked and the NCB subsequently isolated the vacuum brake equipment on the locomotives.

No. 7603 was found to be capable of hauling 2,012 ton trains on the level, 1,116 ton trains on a gradient of 1 in 200, 712 ton trains on 1 in 100, 399 ton trains on 1 in 50, and 268 ton trains on 1 in 33. The train crews found No. 7603 to be deceptively powerful for its size and worked it hard (with the Stephenson, Hawthorn's fitters regularly replacing such items as bearings on site), but found the locomotive to be unstable at speed compared to the 'Big Engines'. Despite this, No. 7603 and her sisters eventually earned the respect of the Whitburn crews who affectionately nicknamed them 'The Stubby Hawthorns'.

No. 7603 spent a period at Boldon Colliery between 21st September, 1961 and 17th November, 1962 and was renumbered No. 6 in 1963. The locomotive was in standard NCB navy blue livery by June 1964 and was again transferred to Boldon on 14th July, 1964.

Engine No.:	4	*Cylinders:*	17 in. x 24 in. inside
Builder:	NER	*Weight:*	67 tons 18 cwt
Works No.:		*Driving wheels:*	5 ft 1 in.
Built:	1883	*Boiler pressure:*	160 psi
Type:	0-6-0	*Fate:*	Scrapped at Whitburn by contractors on 27th June, 1952

The ex-NER '398' class locomotive No. 4 appeared very late on the Marsden Railway scene being transferred from Boldon Colliery in 1950 and was no doubt gratefully received by the train crews in view of her tender cab. In this 29th April, 1952 photograph No. 4 is not actually in steam, the steam is in fact emanating from the massive Whitburn pit boiler in the background. No. 4 was scrapped on 27th June, 1952. *H.C. Casserley*

This 26th August, 1960 photograph shows the tender from ex-NER 'P' class locomotive No. 4 still languishing round the back of Westoe Electric Depot four years after parting company with its engine. The frame was later used to carry a pressurised weedkilling tank. The Westoe winder can be seen against the skyline. *Sydney A. Leleux*

Robert Stephenson & Hawthorn No. 7695 approaches Westoe Lane with the 'Rattler'. The coal dry cleaning plant can be seen in the background, 11th September, 1953. *Alan Snowden*

This locomotive emerged from Gateshead works as No. 1333, one of a batch of 15 class '398' engines built directly by the NER complete with Westinghouse brakes. No. 1333 was rebuilt with a Worsdell steel boiler in August 1901 and was sold to Robert Frazer in 1928 who then sold it onto the HCC. Delivered to Boldon Colliery complete with tender cab, the locomotive became *BOLDON COLLIERY No. 4*.

No. 4 was transferred to Whitburn shed in 1950 to provide emergency cover during the 'Big Engine' crisis.

Engine No.:	7695	*Cylinders:*	18 in. x 24 in. outside
Builder:	R. Stephenson, Hawthorn	*Weight:*	53 tons
Works No.:	7695	*Driving wheels:*	4 ft 0 in.
Built:	1951	*Boiler pressure:*	180 psi
Type:	0-6-0ST	*Fate:*	Scrapped at Boldon Colliery in March 1970

No. 7695 was delivered new to Whitburn Colliery as the second 'mixed traffic' locomotive on 20th December, 1951. The locomotive was transferred to Boldon Colliery on 11th February, 1958 returning to Whitburn on 8th July, 1959. In 1961 No. 7695 was withdrawn from traffic so that its boiler could be fitted to No. 7603, but in time No. 7695 received the overhauled boiler and new fire box that originated from No. 7603 (this taking place in 1963). By this time No. 7695 had become No. 7 and was sporting navy blue livery. In April 1965, the locomotive was transferred to Boldon Colliery.

Engine No.:	7749	*Cylinders:*	18 in. x 24 in. outside
Builder:	R. Stephenson, Hawthorn	*Weight:*	53 tons
Works No.:	7749	*Driving wheels:*	4 ft 0 in.
Built:	1952	*Boiler pressure:*	180 psi
Type:	0-6-0ST	*Fate:*	Scrapped at Whitburn Colliery in April 1968

The third and final 'mixed traffic' locomotive, No. 7749 was delivered to Whitburn Colliery on 17th December, 1952 less than a year before the 'Rattler' service was withdrawn. The locomotive later spent several lengthy periods working from Boldon Colliery, namely 24th July, 1958 to 11th February, 1959, 21st February, 1961 to 5th December, 1962, and 13th May, 1963 to April 1965. No. 7749 was officially renumbered as No. 8 in 1963 (however, the staff at Boldon mistakenly painted No. 9 on the engine). In April 1965, 'No. 9' finally returned to Whitburn shed only to be put into storage pending the fitting of a replacement boiler. The locomotive never worked again.

The crew of Robert Stephenson & Hawthorn No. 7695 take a well-earned break from train engine duties at the north end of Whitburn Colliery station with Whitburn Colliery village forming the backdrop. The date is 29th April, 1952. *H.C. Casserley*

This official Robert Stephenson & Hawthorn works photograph of No. 7749 was taken in 1952 and shows off the powerfull profile of what was described as 'a mixed traffic industrial saddle tank' locomotive. The term 'Harton Unit' was applied to locomotive and wagon alike for allocation purposes by the NCB. *Author's Collection*

Engine No.:	7811	*Cylinders:*	18 in. x 24 in. outside
Builder:	R. Stephenson, Hawthorn	*Weight:*	53 tons
Works No.:	7811	*Driving wheels:*	4 ft 0 in.
Built:	1954	*Boiler pressure:*	180 psi
Type:	0-6-0ST	*Fate:*	Scrapped at Whitburn in April 1968

Bereft of air and vacuum brakes, No. 7811 could be described as a 'freight only' version of the 'mixed traffic' designs. The locomotive was delivered new to Whitburn Colliery on 1st November, 1954 and replaced the last of the 'Big Engines'. After being transferred to Boldon Colliery on 8th July , 1959, No. 7811 was sent to the Lambton engine works for a heavy general overhaul and fitting of a new fire box, returning to Whitburn Colliery on 7th February, 1961. Two further periods were spent away at Boldon Colliery, namely 7th August, 1962 to 10th September, 1962, followed by a lengthy absence from December 1962 to June 1965. While at Boldon, No. 7811 officially became No. 9 (but as this was already being carried by the former No. 7749, the locomotive actually became No. 10!). As with 'No. 9', 'No. 10' returned to Whitburn shed only to be stored pending boiler work (which never actually took place) and when the locomotive was actually withdrawn in April 1968 it was the sole surviving steam locomotive on the Marsden Railway.

Engine No.:	4	*Cylinders:*	18 in. x 24 in. inside
Builder:	NER	*Weight:*	73 tons 14 cwt
Works No.:	28 (1897)	*Driving wheels:*	4 ft 6 in.
Built:	1897	*Boiler pressure:*	170 psi
Type:	0-6-0	*Fate:*	Scrapped at Whitburn Colliery by D.S. Bowran Ltd on 1st August, 1956

No. 1931 was one of a batch of 50 'P' class ('J24' class under the LNER) engines built at Gateshead. It was renumbered No. 5626 by the LNER in March 1946 and was sold directly to the NCB by British Railways in April 1949 finding initial employment at Boldon Colliery. Officially allocated as No. 4 (a number it never actually carried) the locomotive was transferred to Whitburn Colliery on 30th March, 1956 only to be withdrawn from traffic. On 4th April, 1956 the tender was uncoupled from No. 4 and was in fact earmarked to be converted to a weed-killing tank.

Engine No.:	7294	*Cylinders:*	18 in. x 26 in. inside
Builder:	R. Stephenson, Hawthorn	*Weight:*	48 tons 3 cwt
Works No.:	7294	*Driving wheels:*	4 ft 3 in.
Built:	1945	*Boiler pressure:*	170 psi
Type:	0-6-0ST	*Fate:*	Scrapped at Morrison Busty Colliery in July 1974

Built as an Austerity design for the War Department, this engine was delivered as No. WD71485 to the Longmoor Military Railway in September 1945. The NCB purchased the locomotive in April 1947 for use at Boldon

No. 505 busies herself shunting British Railway wagons into the landsale coal depot at Marsden on 16th April,1968. The sidings on the left once formed the spurs into the Marsden quarries with some track lifting in evidence. Today only the house on the right (the former Marsden Post Office) survives.

Ian S. Carr

This 11th April, 1968 photograph was taken from the Marsden Grotto footbridge looking south towards Whitburn with No. 506 (Hunslet 6617 of 1965) toiling through the Grotto Cutting with coal for Westoe and ultimately the Harton Staiths. The trackbed to the Marsden Old Quarry can be seen veering off to the right while the coast road follows the original course of the SSMWCR.

Ian S. Carr

Colliery transferring it up to Whitburn on 11th February, 1959, this becoming the last steam locomotive to be delivered to the Marsden Railway. No. 7294 spent two sojourns away at Boldon Colliery from April 1962 to 10th May, 1962 and from 30th May, 1962 to 22nd November, 1962. The locomotive finally left Whitburn for the Washington 'F' pit on 11th February, 1963.

SSMWCR Diesel Locomotives introduced by the NCB

In 1965 the NCB sent three diesel-hydraulic locomotives to Whitburn shed to take over the remaining duties from the steam fleet.

In contrast to the navy blue steam engines, the diesels carried a light green livery with a darker green band running round the midriff of the engines the width of the inspection doors. Thick red lining was also applied with the lettering 'NCB' appearing in yellow 8 inch letters, supplemented by each number and the wording 'N & D DIVISION' in yellow 4 inch letters while yellow and black 'wasp striping' was applied to the buffer beams.

The diesels were more than strong enough to cope with Marsden mineral duties but were slow compared to the steam engines. With a top speed of 15 mph (rising to 20 mph if running down off Marsden End Bank with a heavy load) they were described by one of the drivers as 'all or nowt'. Prior to this, the steam locomotives regularly attained speeds of 40 mph from Marsden End Bank all the way to South Shields!

Engine No.:	505	*Brake horse power:*	311 @ 1,800 rpm
Builder:	Hunslet	*Weight:*	55 tons
Works No.:	6616	*Driving wheels:*	3 ft 9 in.
Built:	1965	*Fuel capacity:*	225 gallons
Type:	0-6-0	*Fate:*	Scrapped by Booth-Roe Ltd at Rotherham in November 1990

Part of a large divisional order for the NCB, each Hunslet was fitted with a Rolls Royce C85FL 311 brake horse power engine driving a cardan shaft to a Hunslet axle-mounted final drive and reverse gearbox.

No. 505 was delivered to Whitburn Colliery on 31st August, 1965 and worked the Marsden Railway until closure, whereupon it was transferred to Westoe Colliery (in June 1968).

No. 505 then spent a nomadic existence working at several pits in the North-East and in fact became the last of the ex-Marsden locomotives.

Engine No.:	506	*Brake horse power:*	311 @ 1,800 rpm
Builder:	Hunslet	*Weight:*	55 tons
Works No.:	6617	*Driving wheels:*	3 ft 9 in.
Built:	1965	*Fuel capacity:*	225 gallons
Type:	0-6-0	*Fate:*	Scrapped at Lambton engine works in December 1985

Delivered to Whitburn Colliery on 30th September, 1965, No. 506 lasted exactly one week before being withdrawn following an accident at the Bank sidings (which is detailed later). Subsequent repairs saw the locomotive sidelined between 6th October and 12th November, 1965. No. 506 was finally transferred to Boldon Colliery in June 1968.

Engine No.:	507	*Brake horse power:*	311 @ 1,800 rpm
Builder:	Hunslet	*Weight:*	55 tons
Works No.:	6618	*Driving wheels:*	3 ft 9 in.
Built:	1965	*Fuel capacity:*	225 gallons
Type:	0-6-0	*Fate:*	Scrapped by C.F. Booth Ltd at Rotherham in November 1987

No. 507 was delivered to Whitburn Colliery on 5th November, 1965 but spent just six months at the pit before being transferred to Boldon Colliery on 19th May, 1966.

Engine No.:	509	*Brake horse power:*	311 @ 1,800 rpm
Builder:	Andrew Barclay	*Weight:*	55 tons
Works No.:	514	*Driving wheels:*	3 ft 9 in.
Built:	1966	*Fuel capacity:*	225 gallons
Type:	0-6-0	*Fate:*	Stripped for spares and scrapped at Hawthorn Colliery in 1988

This was Andrew Barclay's version of the existing Hunslet diesel-hydraulics at Whitburn and, as such it was fitted with a Rolls Royce C85FL engine but in this case equipped with a Wiseman 15 RLG13-11 final drive gearbox. No. 509 was tested to haul 1,650 ton trains on the level at 15 mph and was delivered to Whitburn Colliery in May 1966. The locomotive was kept busy in the summer of 1966 working 24 hours a day, 7 days a week hauling Harton Group coal down the Marsden Railway for stockpiling at Whitburn due to a seamen's strike. The locomotive was finally transferred away to NCB Springwell Bank Foot engine shed in March 1968.

Miscellaneous Locomotives which appeared on the SSMWCR

An Andrew Barclay 0-4-0 saddle tank locomotive was photographed outside Whitburn shed in August 1934. No details of this rather mysterious engine survive, diminutive by Harton Coal Company standards, it may possibly have been abandoned on the Marsden Railway as a redundant contractor's engine following the completion of the coast road and deviated railway.

In 1943, John Bowes & Partners Ltd loaned the Harton Coal Co. an 0-6-0 pannier tank from the Bowes Railway prior to an intended purchase. The locomotive, *BOWES No. 9* (a Sharp, Stewart design, Works No. 4051 of 1894) underwent working trials on the Marsden Railway between April and July 1943, but was rejected by the HCC due to the apparently high price being asked.

In the early years of Nationalisation, with coal production ever increasing, the crisis regarding the work weary 'Big Engines' was so acute that locomotives had to be hired from the LNER pending the arrival of new designs from the NCB. Two LNER 0-6-0 side tank engines were hired over a period of about four months in 1947. These engines alternated on Marsden duties for a week at a time and originated from Tyne Dock shed. Details remain unrecorded, but the engines were possibly ex-NER 'J72' class side tank engines.

Electric locomotives made daily appearances on the Marsden Railway in the Westoe Lane area as part and parcel of their existing duties, however, two electric locomotives did travel all the way down the Marsden Railway to Whitburn Colliery. No. E10 (a Siemens-Schuckert 4-wheel design of 1913 vintage) was towed to Whitburn shed in 1958 for a repaint due to a heavy work load at Westoe shed. The Whitburn staff mistakenly painted the locomotive the wrong colour and the procedure was never repeated!

No. E6 a Siemens Schuckert B+B articulated design of 1909, was withdrawn in 1962 and towed to Whitburn shed in October of that year to be converted to a diesel locomotive by the shed staff. This proved to be much more complex than originally envisaged and the whole project was eventually abandoned.

Carriages

Some of the locomotives could have been described as work weary when they first arrived on the Marsden Railway. However this condition was as nothing compared to the state of the virtually life-expired carriage stock.

Furthermore the HCC viewed this carriage stock as an irksome necessity and this became more apparent over the years as passenger receipts began to decline. While a policy of high maintenance was applied to the wagon fleet with much repainting and servicing the order of the day, the carriages were simply 'run into the ground' and were seen on the railway in many cases in the liveries and numbering of previous owners. Also, while meticulous records were kept of each and every wagon, absolutely no mention was made of a single carriage in any HCC document.

The reason for this rather odd disparity was quite simple - money. Taking 1920 as an example, a fully loaded coal train running from Whitburn Colliery up to South Shields carried enough coal to earn the HCC an approximate £250 via the Low Staiths, while each 'Rattler' service (if full) running up to South Shields earned the HCC just £4. Little wonder, the carriages came to be viewed with such low esteem!

As with the locomotives, the carriages became something of a menagerie over the years which, nevertheless can be grouped into five separate batches of stock.

First Stock

Nos.:		*Type:*	4-wheel
Builder:	Unknown (ex-NER)	*Compartments:*	4 and 5
Built.:		*Class:*	First and Third
Diagram No.:			

Other details: Non-corridor, elliptical roofs

As can be seen, virtually no details of this initial stock survive except for a couple of early photographs. There were possibly two train formations of 4-compartment carriages with a compartment brake running at each end of the four. The carriages all appear to be various ex-NER designs with one coach at least dating from the late 1860s. All these carriages had elliptical roofs and most had twin running boards. This first stock ran on the railway from the inception of passenger services and were first photographed in 1891. The stock had disappeared by 1919.

Nos.:		*Type:*	4-wheel brake
Builder:	Unknown (ex-NER)	*Compartments:*	3 plus brake
Built.:		*Class:*	First or Third
Diagram No.:			

Other details: Four carriages, non-corridor, elliptical roofs

These were the braked carriages which ran at each end of the two train formations mentioned above. Also of elliptical roof design, they had three compartments for passengers with a separate compartment with double doors and birdcage-type lookout for the guard. These carriages continued in use as 'brakes' for the second stock formations but had disappeared by 1926.

Second Stock

Nos.:	2, 5, 14, 18	*Type:*	4-wheel
Builder:	Ex-Hull & Barnsley Railway	*Compartments:*	5
Built.:	1879	*Class:*	First or Third
Diagram No.:			

Other details: Four carriages. Length 27 ft. Width 8 ft. Wheelbase 16 ft. Height 11 ft 4 in. Elliptical roofs, gas lighting, continuous footboards with individual steps under each door, non-corridor.

These four coaches represented a Kirtley design dating back to 1879 but were in fact built for the Hull & Barnsley Railway in 1885. Prior to World War I they were virtually out of use being employed only on the occasional excursion. Originally air-braked, the coaches had been converted to vacuum brake operation in 1891 but were converted back to air-brake design again in 1918 and were then sent to France to assist in the movement of troops. The War Department repaired and returned the coaches at the end of 1918, and the Hull & Barnsley sold them on to the rolling stock builders and repairers Watts, Hardy & Co. of Howdon, Tyneside for £160 each in 1919. Watts, Hardy then sold the coaches on to the Harton Coal Co. in the same year. These coaches then spent seven years on the 'Rattler' service and were withdrawn in 1926.

Nos.:	58, 59, 64, 71, 72	*Type:*	4-wheel
Builder:	Joseph Wright & Co.	*Compartments:*	5
Built.:	1859	*Class:*	Third

Diagram No.: C4
Other details: Five carriages. Length over headstocks 22 ft 6 in. Body width 7 ft 6 in. Elliptical roofs. Non-corridor.

In all, seven carriages were built under diagram 'C4' by Joseph Wright of Birmingham for the Great North of Scotland Railway. Ordered on 27th May, 1859, the coaches were delivered between 31st July, 1859 and 31st January, 1860. The Great North of Scotland sold five of these coaches to the Harton Coal Co. in March 1920 and they were thus already 60 years old when they arrived on the Marsden Railway. The coaches were withdrawn by the HCC in 1926.

Third Stock

Nos.:	78	*Type:*	4-wheel
Builder:	Metropolitan RCW Co.	*Compartments:*	4
Built.:	1866	*Class:*	Third

Diagram No.: 28
Other details: One carriage. Length over buffers 25 ft 7 in. Width across body 7 ft 9 in. 40 seats, elliptical roof, non-corridor.

No. 78 was one of a batch of just two carriages built for the Great North of Scotland Railway in October 1866. The LNER withdrew the coach on 5th December, 1924 and sold it to the Harton Coal Co. in the same year. Although this carriage was withdrawn by the HCC in 1938, it remained on the system for many years and was seen at Westoe Colliery in 1953 still in its original livery and carrying the number No. 78.

Nos.:	1379	*Type:*	4-wheel
Builder:	Great Eastern Railway	*Compartments:*	5
Built.:	1892	*Class:*	Third

Diagram No.: 402
Other details: One carriage. Length 27 ft. Width 8 ft. Elliptical roof, non-corridor.

The GER built this carriage for its London suburban duties in November 1892, widening it from 8 ft to 9 ft in 1900. The coach was withdrawn by the LNER on 27th December, 1925 and sold to the HCC in 1926. Still in its original livery, No. 1379 even carried a map of the GER (which possibly caused some puzzlement amongst SSMWCR passengers over the years!). Withdrawn by the HCC in 1934, this coach remained abandoned on the railway for several subsequent years.

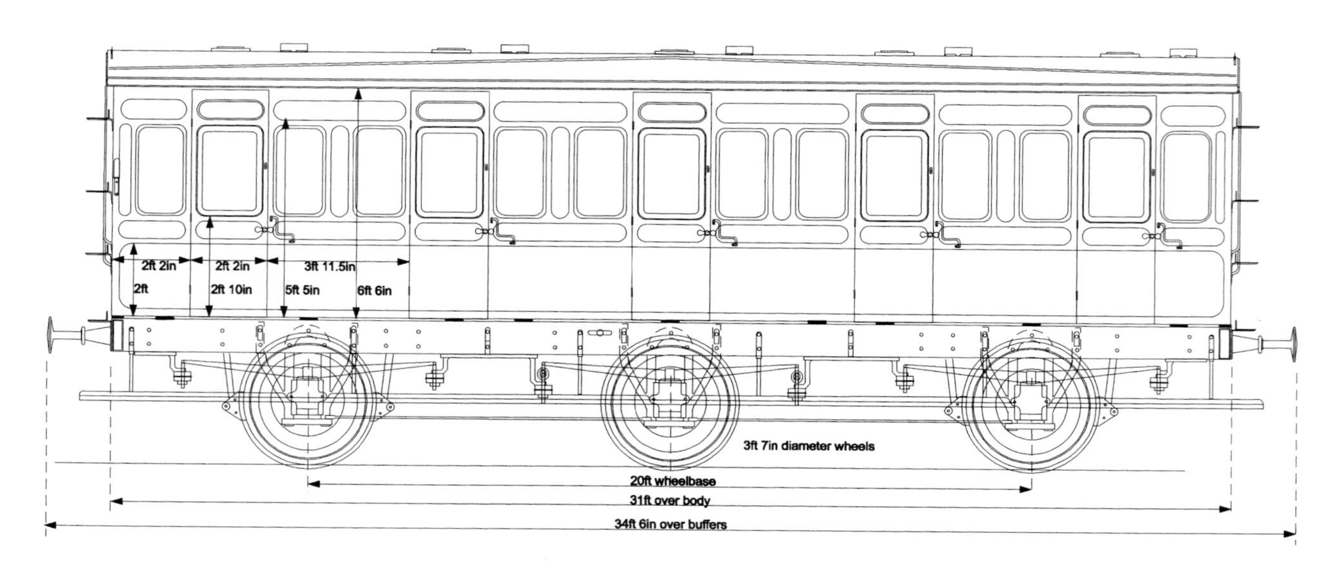

Great North of Scotland Railway Diagram 45

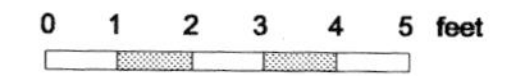

Diagrams 45 pertaining to the GNSR six-wheeled carriages employed on the Marsden Railway.
K. Fenwick

Nos.:		*Type:*	4-wheel brake
Builder:	Great Eastern Railway	*Compartments:*	2 plus brake
Built.:	1883	*Class:*	Third

Diagram No.: 510
Other details: Two carriages. Length 27 ft. Elliptical roofs, non-corridor.

These coaches were two of a batch of 60 built by the GER for suburban use in 1883. The entire batch was withdrawn between 1921 and 1926 with the above coaches sold on to the HCC for third stock braking duties. Apparently these carriages did not carry passengers on the SSMWCR and were possibly converted to carry goods instead as well as performing braking duties. The pair had been withdrawn from 'Rattler' operations in 1938.

Nos.:	83	*Type:*	4-wheel
Builder:	Metropolitan RCW Co.	*Compartments:*	5
Built.:	1865	*Class:*	Third

Diagram No.: 29
Other details: One carriage. 50 seats. Elliptical roof, non-corridor.

This coach was one of a batch of six built for the Great North of Scotland Railway and employed on this system from September 1865 to 3rd December, 1925. Just before it was sold to the HCC, No. 83 earned the distinction of being used in one of the Railway Centenary trains of 1925. The coach had been withdrawn by the HCC by 1938.

Nos.:		*Type:*	6-wheel
Builder:	Unknown (ex-NBR)	*Compartments:*	6
Built.:		*Class:*	Third

Diagram No.:
Other details: Five carriages. Elliptical roofs, non-corridor.

Sadly, the vast majority of North British Railway carriage documentation never reached the respective Record Offices, and the resultant lack of known information is reflected above. Five carriages were purchased by the HCC *circa* 1926 and were withdrawn from 'Rattler' duties in 1938.

Fourth Stock

Nos.:	7200, 7202, 7206, 7209, 7213, 7214	*Type:*	6-wheel
Builder:	Ashbury Rly Carr. & Iron Co.	*Compartments:*	5
Built.:	1893	*Class:*	Third

Diagram No.: 45
Other details: Six carriages. Length 31 ft. Width 8 ft 11 in. Height 12 ft 6½ in. Weight 13 tons 15 cwt. 50 seats. Elliptical roofs, non-corridor.

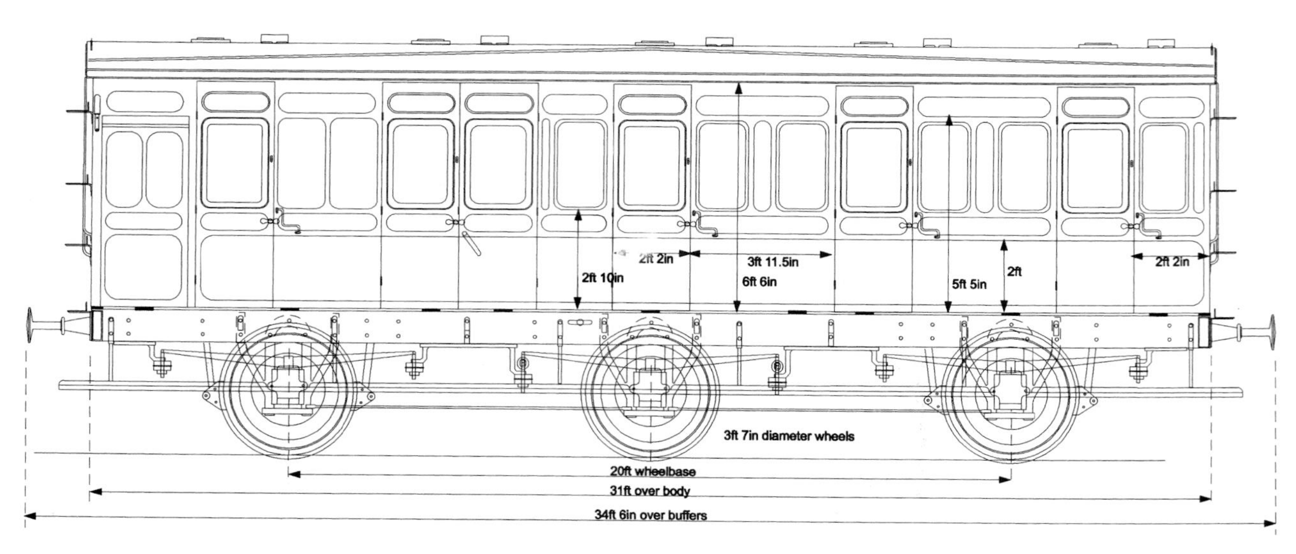

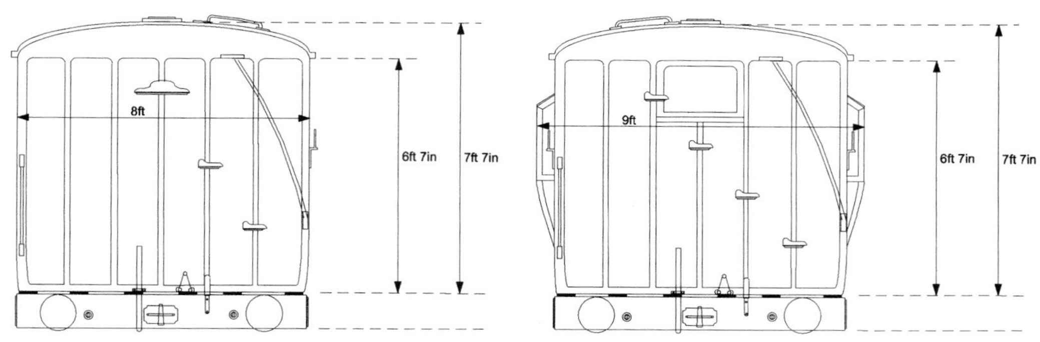

Diagram 46 pertaining to the GNSR six-wheeled brake carriages employed on the Marsden Railway between 1938 and 1953.
K. Fenwick

The NCB lettering really does look out of place in this view of Whitburn Colliery station on 29th April, 1952. This carriage was one of the ex-Great North of Scotland Railway thirds built in 1893. *H.C. Casserley*

These carriages formed an original batch of 78 built for the Great North of Scotland Railway between April and August 1893. The LNER renumbered the coaches by prefixing each number with the numeral '7' and as such the coaches ran in LNER livery. The above coaches were then withdrawn on 19th April, 1937 and sold to Watts, Hardy & Co. who sold them on to the HCC. These six coaches along with the two brakes listed below constituted the 'Rattler' service from 1938 right through until the trains were withdrawn in 1953. As such they came to represent the service at its zenith and (particularly when being hauled by the Pride of the Fleet, No. 8), provided something of an elegant interlude between the seemingly endless streams of mineral trains running along the coast.

Nos.:	7586, 7587	*Type:*	6-wheel brake
Builder:	Great North of Scotland Rly	*Compartments:*	3 plus brake
Built.:	1895	*Class:*	Third

Diagram No.: 46
Other details: Two carriages. Length over body 31 ft. Body width 8 ft (compartment end) and 9 ft (van end). 30 seats. Elliptical roofs. Non-corridor.

Purchased by the HCC in 1938 these two carriages were used for braking duties and as carriages for the general public and white collar workers at Whitburn Colliery. The guard's compartments were equipped with double doors and duckets. The two coaches were withdrawn by the NCB in 1953.

Fifth Stock

Nos.:	NSR 1, NSR 2	*Type:*	6-wheel brake
Builder:	Great Eastern Railway	*Compartments:*	6
Built.:	1892/3	*Class:*	Third
Diagram No.:	404		

Other details: Two carriages. Length 34 ft 6 in. Elliptical roofs. Non-corridor. 60 Seats.

Nos. NSR 1 and NSR 2 began life as suburban coaches numbered 189 and 330, being built in December 1892 and March 1893 respectively. The pair were renumbered by the LNER as No. 60883 and No. 60916 and were withdrawn on 12th June, 1937. The coaches were then subsequently purchased by the North Sunderland Railway where they spent several years plying their way down to the coast at Seahouses. NSR 1 and NSR 2 were then sold to Watts, Hardy & Co. in October 1951 where they were completely gutted (see 'Miscellaneous Carriages' below) before being sold on to the NCB in 1952. The pair were thus to spend just one year on 'Rattler' duty before the service was withdrawn.

Miscellaneous Carriages

The Queen Mother was scheduled to pay a visit to Tyneside in the early 1960s and it was thought that she might visit the Harton system. The NCB realising that they had nothing to convey her in, hastily built a 'Royal Coach' by constructing a carriage body (which resembled a brake van body) and mounting it on a wagon chassis. The Queen Mother never visited in the end, but this inspection saloon was retained for use by any subsequent dignitaries visiting the Harton system.

Although the number of carriages employed in each train formation varied over the years, it was standard practice to couple a third class brake to each end of the 'Rattler' formations.

In later years, just one carriage (usually a brake) was made available for the general public, while all the other carriages were converted for use by the miners. These conversions involved the removal of all the compartments and existing seating from within each carriage and the fitting of longitudinal seating running the full length of each coach. With all platforms lying on the seaward side of the tracks, doors on the landward side were locked, the door handles removed, and the drop-windows boarded up which must have created a strange monocular sensation for the occupants.

After Nationalisation, the lettering 'NCB' appeared on many of the carriages, then in 1953, the last year of services, the NCB suddenly and inexplicably decided to dispense with the existing, and by now faded, previous owner liveries, and repaint the entire carriage fleet a rather anonymous livery of mid-grey with no numbering.

After withdrawal, many coach bodies were removed from their chassis to find further employment as workmen's huts. Three such bodies were employed at Westoe electric depot (two for platelayers, and one as a tool shed) while a fourth body was set down for platelayers alongside Lighthouse signal box.

Wagons

Mineral trains were of course the mainstay of Marsden Railway traffic and the Harton Coal Co. kept meticulous records of their beloved 'big earners'. Originally chaldron wagons were employed on the Harton Railway before being phased out from 1900 onwards. However by 1909 there were still 797 of these wagons to be found throughout the system. At this time 30 chaldron wagons were allocated to Harton Colliery, 12 to Boldon, and 755 between St Hilda and Whitburn Colliery.The chaldron wagons were painted black with white lettering and numerals.

The year 1900 saw the arrival of the first 10½ ton 'P4' class wagons on the Harton Railway, a type which were to become synonymous with the system over the next five decades. These wagons were either built directly by the NER or were built under contract by outside firms such as Watts, Hardy & Co. There was some variations in dimensions and the wagons had cubic capacities which ranged from 355 to 380 cubic feet, tare weights of 5 ton 15 cwt to 6 ton, gross weights of 14 ton to 15 ton, and net weights of 8 ton to 9 ton 7 cwt. The livery was brick red with lettering and numerals in white. Whitburn Colliery was given priority when it came to the allocation of these wagons chiefly because the pit was situated further away from the staiths than the other collieries. The existing chaldron wagons were restricted to a 5 mph maximum speed limit due to their primitive wheel bearings and it was found to be much more economical to employ the 'faster' 'P4' wagons on this 4½ mile journey between Whitburn and the Tyne.

The humble 'P4' 10½ ton wagon catered for all HCC needs and indeed also saw initial service under the NCB. This particular example first appeared on the Marsden Railway in the spring of 1909 and looks capable of many more years' service in this 29th June, 1954 view. The location is Dean Road sidings. *J.P.R. Bennett*

In 1909 with the change-over from chaldron wagon to 'P4' class complete, the total wagon allocation at Whitburn Colliery was as follows :

Nos.	*Weight*	*Type*	*Buffers*	*Notes*
1-6	8 ton	Red Landsale wagons	Dumb	
7-12	10 ton	Red Landsale wagons	Dumb	
13	8 ton	Red Landsale wagon	Dumb	
14-24	10 ton	Red Landsale wagons	Dumb	
25-124	10½ ton	NER 'P4' pattern wagons	Spring	(i)
125	10 ton	GER wagon (side & bottom doors)	Dumb	
126	8 ton	GER wagon (side doors)	Dumb	
127-129	6 ton	Bolster wagons	Spring	
130	8 ton	Low-sided bolster wagon (steel)	Spring	
131-230	10½ ton	NER 'P4' pattern wagons	Spring	(ii)
231-280	8 ton	Ex-Lincoln Wagon Co. wagons	Dumb	(iii)
281-345	10½ ton	New wagons	Spring	(iv)
346-445	10½ ton	NER 'P4' pattern wagons	Spring	(v)

Notes:
(i) Delivered between April 1900 and December 1903.
(ii) Delivered between October and November 1906.
(iii) Delivered April 1908.
(iv) Probably 'P4' wagons. Ex-Frazer & Sons Ltd. Delivered January 1909.
(v) Delivered April to May 1909.

By 1921, the 'P4' wagon stock had reached a peak as far as the Marsden Railway was concerned with 1,151 'P4' wagons allocated to Whitburn Colliery, 470 to Boldon Colliery, and 50 to Harton Colliery.

Thirty years later the NCB announced that the entire 'P4' fleet was scheduled to be withdrawn in favour of 1,028 brand new 21 ton steel-bodied wagons which were required for operations in conjunction with the new coal preparation plant planned at Westoe Colliery. These 21 ton wagons were given a livery described as vermilion with lettering and numerals in white and were ordered from two wagon builders, Charles Roberts Ltd and Hurst, Nelson Ltd. They were delivered to the Harton Railway from 1954 onwards and were not allocated to specific collieries but pooled instead to be worked across the entire system. The 'P4' wagons had become extinct on the railway by 1965.

Miscellaneous Wagons

The 1909 fleet list describes wagons allocated specifically to Whitburn Colliery only, although of course wagons from other pits drifted onto the Marsden Railway from time to time. Six covered-in goods wagons (lettered 'A' to 'F' rather than numbered) were described as 'choppy vans' and were employed throughout the system to deliver choppy (a mixture of horse feed for the ponies working underground) to the various collieries. The quarries at Marsden saw many private owner or main line company wagons arrive at its sidings over the years either for loading with lime dust (in covered wagons) or with limestone (in open wagons) for destinations such as the steel works at Consett. The quarries had their own internal wagon fleet of course and this consisted of 104 iron side-tip wagons employed on the 2 ft gauge systems within the quarries (although officially they were allocated to Whitburn Colliery and carried the numbers 1-104).

One wagon which frequently appeared on the Marsden Railway was the weed-killing tank. This consisted of a pressurised tank mounted on the frames of the former 6-wheel tender of the ex-NER 'P' class locomotive No. 5626.

When maintenance was being carried out on the railway, the platelayers used small hand-worked 4-wheeled trolleys for the carriage of materials along the railway. Great care had to be taken that none of these trolleys were left on the rails after use, and the 'ganger' who was made responsible for them had to ensure they carried danger warnings and were prevented from running away when employed on maintenance duties.

National Telephones: Nos. 333 and 334.
Telegrams: "WATTS, NORTH SHIELDS."

Watts, Hardy & Co.
LIMITED.
BRUNSWICK PLACE, DOCK ROAD, NORTH SHIELDS.

Sole Agents: ROBERT FRAZER & SONS, Ltd., Milburn House, Newcastle-on-Tyne.

RAILWAY WAGON BUILDERS and REPAIRERS

also CONTRACTORS' TIP and LIGHT WAGON BUILDERS of every description for Hire and Sale by immediate or deferred payment. SECOND-HAND WAGONS, WHEELS, and AXLES, and WAGON MATERIAL of all kinds kept in Stock. RE-AXLED WHEELS and AXLES A SPECIALITY.

WAGONS MAINTAINED IN WORKING ORDER BY CONTRACT.
COLLIERY TUBS OF ALL KINDS BUILT TO ORDER.

Advert for wagon builders Watts, Hardy & Co.

The point indicator is set for Wallside traffic in this 1947 photograph which was taken from the Westoe Lane platform. The allotment was tended at one time by the signalman at Westoe Lane. A blast from the approaching 'Rattler' would send him scurrying across the tracks, as he abandoned his gardening tools in favour of point levers. *M.N. Bland*

Chapter Five

Signalling and Safety

When Major General C.S. Hutchinson refused the Whitburn Coal Company's request for Board of Trade sanction of its passenger service in June 1887, one of the reasons given in his report was that there were no proper signalling arrangements at the termini. Taking General Hutchinson somewhat at his word, the WCC duly proceeded to install a signal frame on each terminus platform some time between 20th June, 1887 and 14th February, 1888. Thus began the rather fascinating history of Marsden Railway signalling.

The Railway Signal Company of Fazakerly, Merseyside provided the Harton Coal Co. with all its signalling requirements during the reign of the 'Rattler' service, including these early frames with the necessary equipment.

Westoe Lane signal box was equipped with a 12-lever frame with five levers controlling point work, five levers controlling signals, with the remaining two levers designated as spare. Conversely, Marsden Station ground frame (note that in this case, the official title of the frame actually refers to a station) was equipped with 15 levers with four levers controlling points, eight levers controlling signals, and three levers spare.

The signals themselves consisted of rather lengthy arms mounted on heavy wooden posts with each post crowned by a finial of some elegance. Metal ladders gave the staff access to light the paraffin-fuelled lamp on each arm and these ladders were also used by the train crews on foggy days when the signal aspect could not be seen from the ground! These lower quadrant signals were supplemented by fixed distant signals at two points along the Marsden Railway quite remote from the signal boxes.

The trains were originally controlled along the Marsden Railway by the use of staff and ticket, however the strict rules governing this system of working were apparently being ignored by the Marsden men in these early years. This potentially deadly situation came to the attention of the General Secretary of the Board of Trade on 22nd February, 1900,

> It is also reported to me that there is a very loose and dangerous practice of working trains over the single line without the authorised staff. A case in point is given. The official in charge instructed the driver of a workman's train to travel from South Shields to Marsden without the staff which was at that time at Marsden.

In their defence, the Harton Coal Co. offered the following explanation :

> With reference to the case of a train having travelled from South Shields to Marsden without a staff on 16th February last, we beg to state that this was done under the personal supervision of our traffic manager, and was only undertaken because of an accident to the main line locomotive, which had arrived at Marsden Station with the staff. The staff was taken charge of by the traffic manager at the Marsden Station and the train from South Shields allowed to come through.

The Coal Company then added as a postscript: 'It is not anticipated that such an incident will again occur, as we expect to shortly have working Webb & Thompsons Patent Electrical Staff Apparatus'.

True to their word, the HCC duly installed the Electric Staff machines at both boxes. These Webb & Thompson designs (which were made under licence by the Railway Signal Company) consisted of 3 ft 6 in. tall pedestal machines capable of accommodating 18 staffs at a time in a long vertical slot. The staffs themselves were long metal bars with a ring on each bar. Three dials were built into the instruments directly above the slots. The left dial consisted of a three-position indicator showing UP STAFF OUT, STAFF IN, and DOWN STAFF OUT, the central dial consisted of a two-position needle, while the right-hand dial gave two positions showing FOR STAFF and FOR BELL. Below this third dial was a push-button key for sending bell messages to the other signal box and the signalmen were instructed to give the following bell-coded messages as and when the situation dictated :

Message	*Beats sent*	*How given*
Call attention	1	1
Is line clear?	2	2
Passenger train entering section	3	3
Mineral train entering section	4	4
Train out of section	5	5
Section clear but station blocked	13	3-5-5
Assistant engine in rear of train	4	2-2
Release staff for shunting	7	5-2
Ballast train requiring to stop in section	5	1-2-2
Shunting completed - Staff replaced	7	2-5
Blocking back signal	6	3-3
Obstruction removed	3	2-1
Stop and examine train	7	7
Cancelling signal	8	3-5
Train passed without tail lamp	9	4-5
Testing bells	16	16
Opening and closing	17	7-5-5
Train divided	10	5-5
Time signal	18	8-5-5
Lampman or fog signalman required	19	9-5-5

Whilst on the subject of codes it is pertinent here to list the messages which were to be sent by the various guards to their train crews by whistle:

1 blast - Stop
2 blasts - Go ahead
3 blasts - Reverse
5 blasts - Come to me

As for the drivers, they had to blow their engine whistles twice at the approach to each level crossing.

At about the time that the Electric Staff instruments were introduced, many of the points too remote to be controlled by the signal boxes, but still affecting main line operations, were fitted with Annett's locks. The keys for these locks were attached to the train staffs by brass rings and could thus only be used to release the point levers with the consent of the signalman.

On 20th June, 1903, some resignalling took place in the Marsden station area and the Railway Signal Co. re-locked the Marsden station frame and fitted it with 15 new release labels. Then on 3rd November, 1911, a third signal box was sanctioned by the Board of Trade via the following report:

> Sir,
>
> I have the honour to report for the information for the Board of Trade, that in compliance with the instructions contained in your minute of the 17th October, I have inspected the new works near South Shields on the South Shields, Marsden and Whitburn Colliery Railway.
>
> At Mowbray Road Bridge, which is situated between Marsden and Westoe Lane, a new connection, which is facing to down trains and which leads to sidings has been added on the single line. The points and signals are worked from a new signal box containing a frame of 12 levers of which 4 are spare.
>
> This line is worked on the Electric Staff system, and Mowbray Road Bridge is made into a staff station. It is not, however, a crossing place for passenger trains, and special arrangements are therefore provided to prevent the possibility of trains approaching this box simultaneously from both directions.
>
> The interlocking is correct and the arrangements are satisfactory, so I can recommend the Board of Trade to sanction the new works to be brought into use.

As stated in the above report, this new signal box was situated close to Mowbray Road bridge and was commissioned to protect the main line from the two sets of sidings to the south of Westoe Colliery. The latter was seeing a vast increase in mineral traffic thanks to the newly instigated electrification scheme for the Harton Railway. This connection, which consisted of a single turnout, was situated between Mowbray Road bridge and the new signal box and was facing down traffic.

Mowbray Road Bridge signal box was equipped with a Railway Signal Co. 12-lever frame with locking under the signalman's feet and the levers were fitted with badges rather than release labels. Each lever worked the following equipment:

Lever 1 - Up distant signal.
Lever 2 - Up home signal (main line).
Lever 3 - Spare.
Lever 4 - Spare.
Lever 5 - Up home signal (branch to Colliery sidings).
Lever 6 - Locking bar for point to Colliery sidings.
Lever 7 - Point to Colliery sidings.
Lever 8 - Up home signal protecting main line from Colliery sidings.
Lever 9 - Spare.
Lever 10 - Spare.
Lever 11 - Down home signal.
Lever 12 - Down outer home and distant signal.

From the above lever descriptions, it is clear that Mowbray Road bridge signal box was built simply to oversee the operation of just one turn out illustrating the busy and potentially dangerous nature of this bottleneck.

The year 1920 saw some upgrading of the signalling on the Marsden Railway when Westoe Lane box was equipped with a new Railway Signal Co. 'pit' frame with 15 levers replacing the original 12-lever frame (this being commissioned on 3rd February). At Mowbray Road Bridge signal box, the four spare levers were brought into use to control point work in the sidings alongside the box. Then, on 6th June, the frame at Marsden station was relocked again with the spare levers being brought into use at a cost to the HCC of £1,181 7*s*. 4*d*.

When Marsden station was demolished to make way for the coast road sometime between 1924 and 1929, one of the casualties of the works was Marsden station frame. The railway was then temporarily controlled between Westoe Lane and Mowbray Road bridge by the two existing boxes with one engine in steam operation beyond this on the main line down to Whitburn Colliery. The HCC decided to alter the signalling at the two remaining boxes at this time and upon completion Lieutenant Colonel E.P. Anderson submitted the following report:

Sir,

I have the honour to report for the information of the Minister of Transport that, in accordance with the minute of the 5th November, 1920, I inspected the alterations to the signalling arrangements at Westoe Lane Station on the South Shields Marsden and Whitburn Colliery Light Railway on 28th March, 1929.

The Company's single line runs in a generally east and west direction, Westoe Lane being the terminus as far as passenger working is concerned, though goods trains run to the docks further west.

On the south side of the main line, which runs along the face of the platform, are 3 sidings, connected to the mainline near the east end of the Platform and protected by a trap switch. Another group of sidings situated east of the platform and north of the mainline are similarly connected and trapped. Home signals are provided, for the mainline and for each group of sidings, there is also a mainline starter and fixed distant signals are provided at each end of the station.

The points and signals are worked from a locking frame with 15 levers all in use. The locking was tested and found correct. The distant signals are fitted with red glasses and at my suggestion the Company's representative undertook to consider the substitution of orange glasses for them.

All facing points over which passenger trains run are fitted with locking bolts and bars.

The section from Westoe Lane to Mowbray Road, some quarter mile in length, is worked by electric staff. The number of train movements averages about 30 per day in each direction. Of these about 16 per day in each direction carry Company's employees (free of charge) to and from their work in the Colliery; owing to bus competition there is now practically no other passenger traffic.

At Mowbray Road signal box there are sidings on the northeast side of the line, connected to the mainline by a single set of points facing for trains going towards South Shields. This connection is properly trapped, the points are fitted with locking bolt and bar, and protected by home and fixed distant signals on each side. They are operated from a 12 lever frame, the other levers in which work points in the sidings which are only used by goods trains. The locking was tested as far as the passenger line is concerned and found correct.

The equipment is, generally speaking, suitable for the maximum speed of 20 mph laid down by the Company's rules. But I found at Mowbray Road that the detection of the switch blades by the locking bolt was somewhat slack and that the wheel flanges are striking the sockets which attach the stretcher bars to the switchblades. The Company's representatives undertook to have these matters attended to.

Subject, therefore, to suitable action being taken in this matter, I recommend that approval be given to the use of this new signalling equipment.

I have the honour to be,

Sir,

Your obedient servant

E.P. Anderson Lt. Col.

The one engine in steam operation was found to be a somewhat restrictive practice given the extensive shunting operations in the Marsden Quarries and Whitburn Colliery areas so the HCC built a new signal box opposite the entrance to the two Lighthouse Quarries at Marsden. This was rather an astute move by the Harton Coal Co. as the new signal box came under the Coast Road Agreement and the Coast Road Joint Committee suddenly found themselves paying for a more complex, and thus larger, frame than the one they had demolished at Marsden station!

The Lighthouse signal box was a 25-lever design installed by the Railway Signal Co. at a cost of £1,900 in 1929 with levers 16, 17, and 18 designated as spare. It controlled the main line from the lime kilns down to the Whitburn Colliery station although the point work at the station run-round loop and at the lime kilns low level sidings was released by Annett's key. The signal box was also equipped with a Webb & Thompson Electric Staff machine linked with the existing machine at Mowbray Road Bridge.

Westoe Colliery's dry cleaning plant was constructed in 1933 within the triangle of lines to the east of the pit. New tracks had to be laid to serve this plant and Mowbray Road Bridge signal box was decommissioned and demolished to make way for these. A new box known simply as Mowbray Road was then built 225 yards further down the Marsden Railway well clear of the sidings at Westoe. This was a slightly more complex box than the original one controlling as it did four sets of points rather than one and it thus contained 17 rather than 12 levers although levers 1, 9, and 17 were spare.

In 1935 more upgrading of the signalling took place, with the original 15-lever frame at Westoe Lane box replaced with a new 15-lever frame with all levers working. This was necessary due to the increased traffic flows eminating from Westoe Colliery. Indeed, the station area itself was a busy one at the time and the new frame controlled several safety devices such as trap points and shunting signals. Lever 10 was interlocked with lever 15 and controlled a clearance bar. This bar consisted of a piece of angle-iron which was held level with one of the rails when signal 15 was at danger, dropping beneath the rail level only when the signal was cleared. In this way the clearance bar ensured that any trains waiting to leave Westoe Lane station could not come to a halt too far forward of the platform where they might otherwise have fouled a crossover. Lever 1 controlled a set of points which governed movements on and off the Wall Side Road and Middle Road and here a point indicator (which was worked directly off the point blades) showed the crews which way number 1 point was facing. This looked very much like the existing signals in the area except that it swept through a vertical rather than a horizontal axis. If the points were set for the

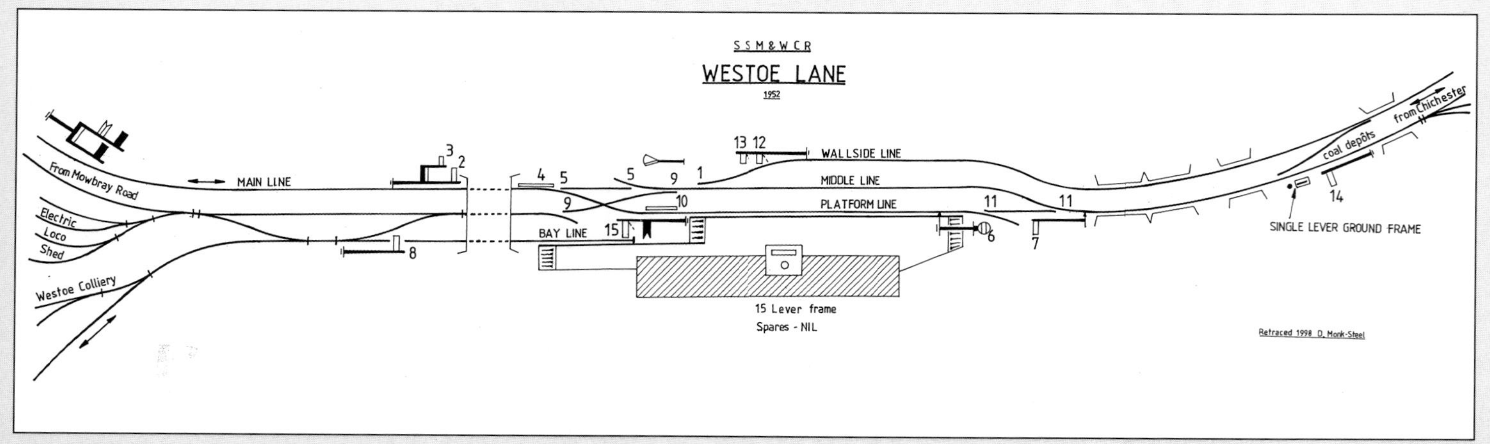

Signalling diagram of Westoe Lane signal box as at 1952. *J.P.R. Bennett/Retraced by D. Monk-Steel*

Signalling diagram of Mowbray Road signal box as at 1952. *J.P.R. Bennett/Retraced by D. Monk-Steel*

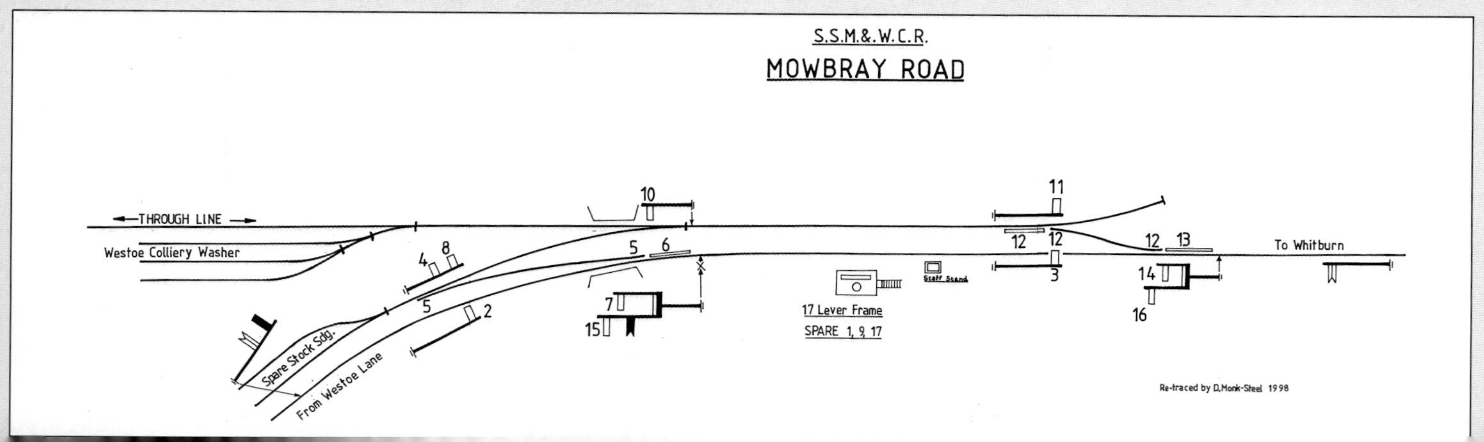

Middle Road, the signal arm was angled in the direction of this Road, but immediately swept over to the Wall Side when the points were changed.

Both Westoe Lane and Lighthouse signal boxes were manned 24 hours a day by signalmen working three shifts, whereas Mowbray Road box was only manned between 6.00 am and 10.00 pm. Between these times there was a long single section between Westoe Lane and Lighthouse boxes but otherwise the staffs were exchanged at Mowbray Road. When the 'Rattler' passed Lighthouse box with a service to Whitburn Colliery, the staff was swapped with an Annett's key so that the train crew could release the points for the run-round-loop at the station. On the return journey, the crew collected the staff and threw the key onto a tray next to the signal box window. The Lighthouse signalman was furious if the crew missed because he then had to walk down to the track to retrieve the key.

Generally the signalling arrangements ran well and the passenger service ran to time, however, on one occasion the 'Rattler' appeared to have provided an express service. Lighthouse box was shut on Sundays and one particular Monday morning the 'Rattler' was ready to leave Whitburn Colliery station with the first train of the day, the 6.00 am. The signalman was late on this particular morning but the train crew found an off-duty signalman instead who opened up the box and released the train in time for its booked departure. The duty signalman finally arrived at 6.09 am and booked not only his start time as such but also entered the train departure time as 6.09 am so that his workmates would not get into trouble for releasing the train 'unofficially'. The 'Rattler' had long gone by this time and was in fact making its final approach to Westoe Lane station with the Westoe Lane signalman duly booking its arrival at the station as 6.12 am. All seemed well until a few weeks later when the traffic management duly hauled up a somewhat puzzled train crew before them and demanded to know why it had only taken them three minutes to travel from Lighthouse to Westoe Lane (given the start-stop times this would have required speeds in excess of 90 mph!).

Given the extravagant signalling on the Marsden Railway (block working, electric train staffs, *and* one engine in steam) serious collisions and accidents were thankfully avoided over the years, however lesser accidents occurred as did a curiously high proportion of deaths which were due in some cases either to suicide (the victims seeming to prefer the cold finality of a locomotive wheel to a 100 feet plunge off the Marsden Cliff) or to, it has to be said, drunkenness.

However, the very first recorded fatal accident featured neither of these and occurred in 1891. On 7th July of this year, Robert Smith, a Marsden Railway engine driver was killed. The whistle on his engine had jammed open and this must have irritated him to such an extent that he felt the need to clamber up onto the roof of the moving engine in order to free the whistle.

Tragically the engine passed beneath Trow Rocks footbridge as the unfortunate Smith was engrossed in this running repair with the inevitable consequences.

The events of 5th October, 1892 must have returned to haunt Christopher Robinson time and again over the years. Driver Robinson was taking the 5.40 am 'Rattler' service down the coast to Marsden on this Saturday morning when he noticed on the approach to Bents Cottages what appeared to be a 'bundle'

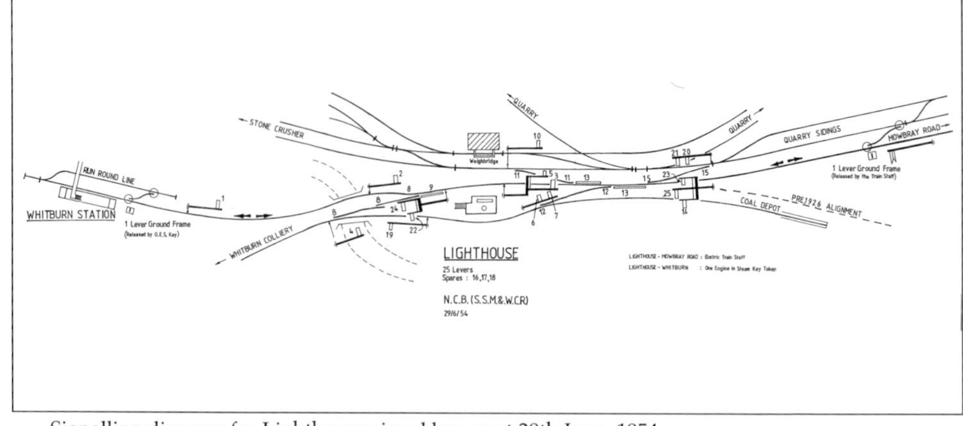

Signalling diagram for Lighthouse signal box as at 29th June, 1954.

J.P.R. Bennett/Retraced by D. Monk-Steel

The interior of Lighthouse signal box on 12th March, 1957 with the road set for train movements between Whitburn Colliery sidings and the Quarry sidings. Note the operations board behind the levers and the electric staff machine complete with staffs at the far end of the box.

I. Scrimgeour

lying between the tracks. Robinson immediately applied his brakes and then began furiously blowing the whistle as he realised that the 'bundle' was in fact a woman holding a struggling child. Robinson's prompt action brought the train to a halt just short of the woman and child. Both were soaked to the skin and when questioned later it was found that the woman was deeply disturbed and had been trying to drown herself and the child in the sea prior to approaching the railway.

Richard Newton was one of a small gang of Whitburn miners who left the Marsden Inn late one Saturday night back in April 1894. Trudging down the footpath which lead to Marsden, the men suddenly decided to climb up over the railway rather than walk down to the little underbridge. As they were about to step across the rails they saw the lights of the 11.10 pm 'Rattler' from Westoe Lane approaching fast. All stood back save for Newton who, shrugging off the restraining arms of this workmates, shouted 'I'll get clear'. They were the last words he spoke.

On 13th December, 1896, the guard of the 6.15 pm service from Westoe Lane gave the following statement to the Police:

> We left Westoe Lane behind number 3 with William Newton in charge. I heard the engine whistle blow as we were approaching Trow Rocks Bridge. It was usual to blow before we got to it. We went a little further when I heard the whistle blow sharp as though it were an alarm. I was then jerked as though the train had suddenly stopped. I got out of my van and went back and found the deceased lying at the side of the railway.

The deceased turned out to be a possible suicide victim as he had stepped purposefully in front of the engine.

On 25th November, 1899, John Millet brought his engine back 'light' after working the 9.30 pm service down to Marsden. He left the engine in a siding near Westoe Lane, signed off duty, and was walking home down Sphor Terrace when he heard the whistle of his engine being blown furiously. His fireman was responsible for using the whistle after discovering the body of Michael Dunleary lying across the tracks. Dunleary having missed the last train had been walking down the tracks to his home at Salmon's Hall possibly assuming that all train movements on the railway had ceased for the night. It was an assumption that cost him his life.

In June 1901, a train of 20 empty wagons arrived at the quarry sidings off Lighthouse bridge. The guard jumped down from the train and studied the small yard to see where the train was to be shunted. Choosing a siding containing six box wagons and two mineral trucks, the guard uncoupled the train from the engine and allowed this train to run by gravity down into the siding by releasing the brakes. He then repeated this fly shunt by running the accumulated wagons further into the siding. Colin Atkinson, a greaser on the railway, had been resting on one of the two mineral trucks with one leg dangling in front of the buffers. The leg had become badly crushed in the fly shunt and within 10 minutes Atkinson was dead.

On 7th October of the same year, another employee suffered an accident on the railway, thankfully in this instance without fatal consequences. On this Saturday night, the gentleman had entered Westoe Lane station the worse for

Lighthouse signal box was an immense structure by industrial railway standards but was virtually redundant by the time this 29th June, 1954 photograph was taken. The wagons in the distance stand outside the landsale coal depot. *J.P.R. Bennett*

No. 505 heads a train of British Railways 21 ton wagons toward the redundant Lighthouse signal box on 16th April, 1968. The summit of the SSMWCR (at 119 feet above sea level) lay between the signal box and Lighthouse bridge which can be seen in the distance. *Ian S. Carr*

drink with the intention of walking the railway down to his home at Marsden. As he passed a locomotive standing at the platform, he staggered, fell against it and slipped down between engine and platform. He became entangled with the brake equipment and the driver had to dismantle parts of the engine to free him from it. The unfortunate man suffered scalp and leg injuries.

On 6th October, 1903, William Hunter a stoneman at the Marsden Quarries left Smith's Farm for his home at Marsden. On his journey home he sat down for a drink of whiskey and must have fallen asleep. Unfortunately he was so drunk that he had not realised that he had chosen to sleep with his legs across the metals of the Marsden Railway. The 9.30 pm from Westoe Lane duly made its way down the railway with the inevitable and fatal consequences.

James Graham was one of a crowd of Whitburn miners who were standing on Westoe Lane station awaiting the arrival of the 4.55 am service down to the pit on 10th June, 1909. As the empty train was being backed in to the station, Graham, who was at the eastern end of the platform, attempted to board the moving carriages. He missed the footboard, slipped down between the carriage and platform, and was dragged 50 yards receiving severe and frightful lacerations to his back.

On 16th November, 1909, Thomas Morrison the Marsden station master was standing on his platform awaiting the arrival of the 'Rattler' from Westoe Lane when to his horror, he saw a runaway horse and cart clattering along the main line with several people giving chase! The owner of the offending vehicle, William Stott was meanwhile blissfully unaware of the situation and was enjoying a drink at Marsden Inn. Thankfully the offending obstacle was safely removed from the railway in time but Stott was charged with being an improper distance away from his horse and cart and was fined 20 shillings plus costs.

George Gough was taking his train down to Marsden on 11th February, 1910 when he noticed something lying next to the line. On the return journey, Gough brought his train to a halt to discover that the 'something' was in fact William Douglas, a gentleman who must have been blessed with a strong constitution. Douglas had suffered a severe head injury after being struck by a previous train, was covered in frost, but had survived.

An accident occurred on 16th January, 1911 that caused neither death nor injury, but which was ironically the most worrying for the Harton Coal Co. as the accident could not be put down to human error. An evening coal train consisting of eight chaldron wagons sandwiched between two rakes of 10½ ton wagons was making its way up the Marsden Railway towards Westoe. The train had just crossed Mowbray Road bridge when the leading chaldron wagon became derailed, broke free from its 10½ ton counterpart, and sent itself and the seven other chaldrons hurtling down the embankment. Not content with this, the wagons proceeded to dash an electric lamp standard to the ground, before settling themselves rather neatly on top of each other, just where, until that moment, a greenhouse had been standing! Each wagon had been holding some 3½ tons of coal and this now lay scattered across several gardens much to the glee of the local residents. The Harton Coal Co. wagon lists described the chaldrons as being 'broken to atoms' in the accident which was blamed on a broken axle on the leading wagon. This derailment heralded the start of the wholesale withdrawal of the HCC chaldron wagons, and by September the fleet had been reduced from 785 to 615 examples.

After spending several hours in Marsden Inn one December day in 1914, Alexander Richardson and his brother began to make their way home to Marsden. Thick fog and the effects of alcohol caused the brothers to lose their way with the men eventually discovering not the footpath but the Marsden Railway. Both men climbed the fence near Redwell Lane bridge and began to walk the track down to Marsden. Meanwhile driver Henry Haswell had already started his train from Westoe Lane at the booked time of 9.30 pm and was on his way to Marsden. Alexander Richardson was run down and lost both legs. He died later.

Early on the morning of 31st August, 1927, Alfred Bulmer brought his train to a halt between Salmon's Hall crossing and Frenchman's crossing after discovering a suicide victim's decapitated body lying against the tracks. This was the last accident officially recorded in the Harton Coal Co. reports. One further recorded accident took place however, (happily without fatal consequences) when the Marsden Railway was under National Coal Board ownership.

On 6th October, 1965, Whitburn Colliery sidings were a busy and indeed fascinating place with steam locomotives still going about their duties while the Hunslet diesels were being commissioned. On this particular morning an immaculate Hunslet No. 506 was standing at the head of a train of coal trucks while a Hunslet representative busied himself preparing the locomotive for driver training. The train duly moved off into the thick morning sea fog with both representative and trainee crew on board, however another Hunslet, a steam version, No. 9 (3191 of 1944) had other ideas. This was leaving the pit with a long train of empties bound for the Marsden Quarries and the two trains collided head-on at slow speed. In the aftermath it was found that the steam locomotive had suffered a slight crumpling to the buffer beam which required straightening in the workshops. The diesel however, still resplendent in its immaculate livery, stood totally unmarked and the Hunslet representative naturally seized upon this opportunity to boast about how robust these diesels were in comparison to the now outdated steam engines. The representative then attempted to restart the engine without success, the inspection doors were duly opened and it was found that the engine had been torn from its frame mounting bolts and been shoved back several inches. The engine was in fact written-off side-lining the diesel for six weeks while a new engine was fitted. As for the steam locomotive, this simply continued about its shunting duties.

After the withdrawal of the Marsden Railway passenger services in 1953, the previously strict signal arrangements were relaxed somewhat. Only the dwindling mineral trains plied their way along the branch now so the block signalling was seen as an extravagance. By 1957 the electric train staff instruments were taken out of use and replaced with single line staffs operated manually. Communication between the signal boxes was abandoned shortly afterwards while the signal boxes themselves eventually became glorified ground frames, being used by the train crews when points needed changing. The signal boxes in this capacity still stood guard over the railway until it finally closed.

Chapter Six

Heavy Traffic on a Light Railway

The following account is designed to illustrate 'a day in the life' of the Marsden Railway. Taken from the short but fascinating period between the end of World War II and the withdrawal of passenger services in 1953, this snapshot view shows a little branch line being worked very hard indeed.

At the time, four engines were in steam on any given working day, however as can be seen, the train engine took the lion's share of the work with the other three engines taking on supporting roles. The 'Rattler' was run day and night at often curious times in order to cater for the complex and overlapping matrix of shift patterns at Whitburn Colliery and this service had to be threaded between the heavy movements of coal.

6.00 am. The day shift begins with two engines in steam, the train engine and the early pilot. The last duties of the previous night shift drivers would have been to ensure that the engines had sufficient supplies of coal and water to perform initial workings, while the firemen ensured that the engines were equipped with shunting poles. The day shift drivers, after signing on, checked that their locomotives were equipped with the following items:

One large lamp. Two side lamps. One hand signal lamp capable of showing red and white lights. One water gauge lamp.
One hand hammer.
One complete set of oil cans.
A quantity of flax, spun yarn, and twine. Red flags.
A sufficient supply of fuel and water.

All shift change-overs took place at Whitburn Colliery (with regards to the Marsden Railway operations) and at 6.00 am the train engine took the first 'Rattler' service of the day up to Westoe Lane leaving the early pilot to shunt wagons between the Bank sidings under the screens and the storage sidings.

6.20 am. Having run round its train at Westoe Lane station, the train engine departs with a service for Whitburn Colliery, its passengers consisting mostly of pit mechanics. On arrival at Whitburn, the train engine is uncoupled and 'goes for bunkers', in other words it is taken down to Whitburn shed to take on coal and water. Here the driver had to turn a large wheel to release the water from the large storage tanks at the shed while the firemen and guard took turns to shovel coal into the tender. The 'Big Engines' consumed between 2 and 4 tons of coal per shift and were treated to top grade Whitburn steam coal (reckoned to be the best in the North-East) and this must have contributed greatly to the efficient running of the train engine over its long and arduous duties. Alongside the big train engine would be the diminutive quarry engine, 'the bunkering' of this latter locomotive being a far easier task for her crew.

7.00 am. The quarry engine scuttles off towards Marsden to commence her daily sojourn at the quarries while the train engine is reunited with its carriages at Whitburn Colliery station.

A busy scene at Westoe Lane station. Harton electric locomotive No. 12 can be seen (*left*), with No. 11 in the foreground. The 'Rattler' rolling stock stands in the platform road. *Alan Snowden*

The date is 29th April, 1952 and English Electric No. 12 glides through the Middle Road at Westoe Lane with a mineral train for Westoe Colliery while Robert Stephenson & Hawthorn No. 7695 waits to leave the station with a 'Rattler' service for Whitburn Colliery. There were two remarkable similarities between these two very different forms of motive power, both weighed 50 tons and both were built in 1951. *H.C. Casserley*

7.15 am. The 'Rattler' leaves Whitburn with a train load of stone men and other shifters many of whom alight at Marsden Cottage Halt. On arrival at Westoe Lane, the carriages are left at the platform while the train engine moves off towards Westoe Colliery in order to pick up a train of empty wagons which has been assembled by one of the electric locomotives. The train engine then takes the empties down the line to Whitburn Colliery, deposits them in the pit sidings, and collects a loaded train from the Bank sidings under the screens.

The freshly mined coal heaped up in these wagons has itself taken a wholly separate and fascinating journey. Cut from the face by the hewers, it is broken up, loaded into mine tubs, and horsedrawn to the main haulages. Here it is tipped into larger tubs and rope-hauled for several miles to the shaft bottom. Once at the surface, the tubs are run onto a circuitous track which feeds them onto a weighbridge so that each miner's quota can be calculated and recorded. Each tub is then fed into a tippler which flips it over allowing the coal to teem down onto the jigging screens. These screens consisted of sloping, shaking tables with holes of ever increasing sizes so that the coal is separated into three grades (small coals, cobbles, and large coals) with each grade fed onto separate shutes positioned above the various railway wagons. A separate shute fed splints (the lowest grade) into a wagon which collected this for the miners' free coal allowance. However, unbeknown to management, the top grade coal somehow always managed to find its way down this shute instead!

As for the 'dirt' or stonewaste, the bulk of this was separated underground with the remainder hand-picked out at the screens for loading into further separate railway wagons. No water was used in the screening process and in the dark and windowless confines of the screenhouse, the dust became virtually intolerable. On particularly bad days during a hot summer, the screenmen would ask for a dust bonus. The foreman would then be called for, and he would hold his hand up just six inches from his face. If he could see his hand, the dust bonus (which amounted to just 4*d*. a day) would be denied!

The train engine duly hauls the loaded coal train out from under the screens and runs it up the line to Westoe Colliery sidings so that the duty electric locomotive can break it up and propel each section into the dry cleaner for further refining. This dry cleaner consisted of a 90 ft high tower which was worked on a process similar to the vacuum cleaner. The driver had to ensure that no-one was standing in the way as he propelled the wagons into the cleaner, but the dust was so intense that he could not even see the wagons in front of him! Leaving this unenviable task to the duty electric locomotive, the train engine returns to its carriages at Westoe Lane station. The firemen cleans the fire, the engine is watered, and the crew then have their breakfast.

9.00 am. The late pilot engine leaves Whitburn shed to begin its day's work assisting the early pilot with shunting duties around Whitburn Colliery and the railway now sees its allotted four locomotives all in steam.

9.30 am. The train engine leaves Westoe Lane with a train load of back shift (or middle day) miners and takes them down to Whitburn Colliery. Here the engine is detached and run light up to the colliery to collect another loaded coal train from under the screens. This coal train is then taken up the line and deposited in the sidings at Westoe Colliery. Meanwhile, one of the Whitburn pilot engines busies itself marshalling both HCC and private owner wagon loads of limestone and lime dust which have been brought out of the quarries by the quarry engine. The pilot then hauls these wagons down from the crushing plant sidings to Whitburn Colliery sidings where they are coupled up to covered wagons which contain either items from the pit stores for the other collieries, or 'choppy'. At the other end of the branch at Westoe Colliery, the train engine has been coupled up to a rake of empty coal wagons and it hauls these down to Whitburn Colliery.

11.00 am. The train engine leaves Whitburn with 'The Goods', the aforementioned train of stores and quarry wagons, and takes these up the line, not to Westoe Colliery, but to the wallside, the track at Westoe Lane station furthest from the platform. The train is brought to a halt here so that the staff in the station's toll office can record the weights of lime and limestone in the private owner wagons. Meanwhile, the train engine is uncoupled in favour of electric traction so that the 'goods' can be taken down Chichester Road Bank to Dean Road sidings. Here, the private owner wagons were separated for collection by main line locomotive, their final destination being in many cases the steel works at Consett.

Back at Westoe Lane station, the uncoupled train engine runs light over to the sidings at Westoe and collects another train for Whitburn. One of the few loaded trains to run down to Whitburn, these wagons held either stonewaste from the various Harton pits, or boiler coal which had been refined through the dry cleaning process at Westoe. On arrival at Whitburn, the train engine is detached and sent 'for bunkers', while a pilot engine takes the stone waste down to one of the two former Harbour quarries for tipping and then returns to move the boiler coal down to the colliery boiler house which was adjacent to the shed. Meanwhile, its tender brimming with coal and water for the heavy 'Rattler' schedule ahead, the train engine returns light to Whitburn Colliery platform to be reunited with its carriages.

12.10 pm. The train engine leaves Whitburn Colliery with a 'Rattler' service for Westoe Lane. On board are mostly boilermen, boiler firemen, and subsidiaries (the name given to the Marsden quarrymen).

12.30 pm. Having run round its train at Westoe Lane, the train engine sets off with another service full of afternoon shift miners and bankhands down to Whitburn Colliery.

1.00 pm. Another quick run-round sees the train engine hauling the 1.00 pm 'Rattler' away from Whitburn Colliery, this time taking many day shift miners home (these miners alighting at either Marsden Cottage Halt or Westoe Lane). Water is taken on before the next service leaves the station.

1.30 pm. The 'Rattler' leaves Westoe Lane, its carriages full of screen workers and at Whitburn Colliery the train crew run the engine round before walking down to Whitburn shed to book off duty.

2.00 pm. The afternoon shift begins with a 'Rattler' service departure from Whitburn Colliery full of day shift miners. After running round at Westoe Lane, the train engine is again watered.

2.30 pm. The 'Rattler' leaves Westoe Lane with the 2.30 pm to Whitburn Colliery. However this was a balancing service to get the carriages down to Whitburn in order to collect the screenlads and bankhands at the end of their shift, and on many occasions this train ran empty. On arrival at Whitburn, the train engine is detached and sent 'for bunkers'.

3.00 pm. While the train is being coal and watered, the early pilot engine comes onto shed at the end of its duties leaving the train engine, the late pilot, and the quarry engine still in steam. Sufficiently fuelled up, the train engine returns to the carriages standing at Whitburn Colliery platform.

3.15 pm. The 'Rattler' leaves Whitburn with the aforementioned screenlads and bankhands safely on board.

3.30 pm. After a quick run-round the train engine is ready to leave Westoe Lane with another balancing working for Whitburn. On arrival at Whitburn, the carriages are again left at the station while the engine runs light to the screens to collect a loaded coal train. This is taken up to Westoe sidings, deposited, and empties then collected for the return to Whitburn. At Whitburn pit the train engine is uncoupled and run light to its waiting carriages at Whitburn Colliery platform.

4.00 pm. Its daily nine hours of duty completed, the quarry engine runs down to the shed at Whitburn.

4.50 pm. The train engine takes a train (of mechanics) to Westoe Lane.

5.05 pm. The 'Rattler' (on another balancing working) is brought back down the line to Whitburn. The train engine is then run light to the shed to be topped up with 'bunkers'. Here the crew busy themselves cleaning out the ash pan and smokebox, and oiling round the engine before taking the locomotive back to Whitburn Colliery platform. After coupling the engine to the carriages, the crew have tea and sandwiches.

6.25 pm. The 'Rattler' is run up to Westoe Lane with many back shift miners on board. The train engine is detached and run light to Westoe sidings to collect empty wagons for Whitburn. These are duly deposited at Whitburn Colliery and a loaded coal train collected for a run back up the line to Westoe Colliery. The train engine is then returned to Westoe Lane, coupled to the 'Rattler', and watered.

7.00 pm. The late pilot engine completes its last shunting duties around Whitburn Colliery and runs down to the shed leaving just the train engine in steam and thus responsible for all non-electric overnight railway movements.

7.50 pm. The 'Rattler' leaves Westoe Lane with night shift drillers and cutters on board. On arrival at Whitburn, the train engine is detached and run down to the screens to collect a loaded coal train which is taken to Westoe. It then returns down the branch with empties for Whitburn which are duly deposited under the screens, before the train engine is run back to Whitburn Colliery platform and the 'Rattler'.

9.25 pm. The 'Rattler' leaves Whitburn with afternoon shift miners and bankhands on board.

9.40 pm. The train engine returns from Westoe Lane with another 'Rattler' service, full this time with pit stonemen and shifters. On arrival at Whitburn Colliery, the train crew walk down to the shed, their afternoon shift complete.

10.00 pm. The night shift begins. A loaded coal train is collected from the screens at Whitburn and taken up the branch to Westoe Colliery. The train engine then collects a train from the sidings at Westoe which consist of both empty wagons, and wagons loaded with boiler coal. This train is then hauled down to the boiler house sidings at Whitburn Colliery and detached while the engine 'goes for bunkers'. The engine then collects wagon loads of boiler ash from the boiler unloading point and runs these down to one of the two waste tips alongside the pit. The engine then returns light and propels the wagon

A permanent way man can be seen greasing point mechanisms (*right*) while a 'Stubby Hawthorn' takes water at Westoe Lane. *Alan Snowden*

An unidentified 'Stubby Hawthorn' nudges up to the water column at Westoe Lane station on 3rd March, 1952. The original platform continues up to Ladysmith Street footbridge while beyond, a brace of electric locomotives stand outside Westoe Electric Depot. *M.N. Bland*

loads of fresh boiler coal into the boiler area loading point before returning light engine to the platform at Whitburn Colliery.

11.30 pm. The train engine heads a 'Rattler' service up to Westoe Lane (another balancing working, this train also was nearly always empty). The engine then runs light to Westoe sidings to collect a train of wagons loaded with stonewaste, and these it takes directly to one of the tips at Whitburn. The train engine then runs light all the way from Whitburn Colliery to Westoe Lane station and this was the only such movement performed by the locomotive over the entire 24 hour period, illustrating just how efficiently the Marsden Railway was run at this time.

1.00 am. The train engine leaves Westoe Lane with a 'Rattler' service full of early shift coal fillers. On arrival at Whitburn Colliery, the engine is run round ready for an 1.30 am departure.

1.30 am. The 'Rattler' is hauled up the branch to Westoe Lane as a balancing service, and, on arrival at the station, the engine is run round and the train crew enter the station buildings for well earned tea and sandwiches.

As the train engine is left to simmer at the station a sudden stillness descends over the Marsden Railway with no booked steam movement scheduled for the next three hours. Only at Whitburn shed is any railway activity evident, with spare engines being washed out or repaired. At 2.00 am the fire is lit in the early pilot engine, followed by the quarry engine at 3.00 am, and the late pilot engine at 5.00 am.

4.30 am. The train engine leaves Westoe Lane with the 'Rattler' service taking dayshift miners and pit boiler men down to Whitburn for the start of their shift.

5.00 am. After being run round, the train engine departs from Whitburn Colliery platform with another 'Rattler' working, this time taking night shift drillers home to South Shields.

5.30 am. The train engine departs from Westoe Lane station with second shift miners on board the 'Rattler'. This was the final booked working for the nightshift crew and, as one of the firemen who regularly worked this shift lived at Horsley Hill, his driver used to drop him off at Marsden Cottage Halt to save him the long walk back home from Whitburn. This left the driver in sole control of the locomotive on its journey down to Whitburn. After uncoupling the engine at Whitburn Colliery platform, the driver had to path it through a maze of point work down to the shed. Several spring points (installed to ensure the safe passage of runaway wagons) barred the way to the shed and here the driver had to jump from the moving engine, run forward, hold the point open while the engine passed through, then give chase in order to reach the footplate again. With the tank engines this was simply a case of 'a hop, skip, and jump' to get back into the cab. However, the tender engines were quite another affair, and here the driver had to stand and watch while the long 6-wheel tender passed safely through the points before giving chase down the full length of the tender in order to reach the cab. On the falling gradient the engine would pick up speed quite alarmingly, on occasion forcing the driver to into a full blooded sprint!

With the train engine fuelled and safely back at Whitburn Colliery platform, the night shift duties ceased with the day shift crew ready to take over for their 6.00 am departure.

The vast majority of the Marsden steam fleet was scrapped round the back of Whitburn shed over the years and even the Pride of the Fleet was not to escape this sad demise as this 29th June, 1954 photograph clearly shows. The final ex-NER tender engine on the railway was scrapped in 1956. *J.P.R. Bennett*

Even after the 'Rattler' had ceased to exist, Westoe Lane continued to be a place of interest as this 12th March, 1957 photograph clearly shows. The station signal box is still in use (with duty signalman just discernible) controlling in this case, two electric locomotives through the Westoe Lane bottleneck. A German-built locomotive of Edwardian vintage is standing at the platform while an English Electric version of 1951 glides down the Wallside road. *I. Scrimgeour*

Chapter Seven

The Final Years

The withdrawal of the 'Marsden Rattler' was the first of several changes taking place in the 1950s which came to alter the entire face of the railway once again.

The reason for the 'Rattler's' demise, the Westoe washer, went into full scale production in 1954 and with it came the cavernous 21 ton steel-bodied wagons. These vermilion wagons heralded the wholesale and sudden withdrawal of the brick-red 10½ ton wagons of NER vintage, which for so long had been a feature of the Harton Railway. With these little wooden-bodied wagons went the last of the NER 'Big' engines, themselves replaced by a fleet of sturdy tank designs. Empty wagons were still hauled 36 at a time, but full train lengths were reduced from 36 wagons down to 18 wagons. Even the coal heaped up in these new wagons looked different. With the screenmen now relying on the immaculate cleaning and grading properties of the Westoe washer, the mineral at the Whitburn screens was simply dumped in the wagons regardless of size or quality.

With the passenger trains gone much of the trackside architecture became superfluous. However, even though defunct such architecture remained in place as shadowy reminders of the days of the 'Rattler'. Westoe Lane station, its platform side track now equipped with overhead wire, was still employed for office use. The platform at Marsden Cottage Halt although bereft of its little shelters could still be discerned amongst the encroaching undergrowth, while the platform at Whitburn Colliery enjoyed a similar survival.

For the signal boxes, the loss of the passenger timetable meant a long slow decline. The electric staff apparatus was dismantled in each box to be replaced by staff and ticket operation (though even this method was itself eventually abandoned in favour of one engine in steam). Dismantling of the interlocking and signals also took place in the ensuing years with the signal boxes used by the train crews for point changing only. This said, the boxes remained *in situ* right through until closure of the railway itself, while even the signals themselves remained in place. Ignored by the train crews, these signals stood as forgotten and rusting sentinels over the railway.

As can be seen much of interest was lost during the 1950s, however the Marsden Railway had not so much been downgraded as simply returned back to its original role of mineral railway. Whitburn Colliery was producing as much coal as ever and by 1956 had become the biggest coal producer in the group of five Harton pits. All this was to change however with the announcement in 1957 that Westoe Colliery was to be the recipient of a 10 year reconstruction which was to see it upgraded to the status of 'super pit'.

Meanwhile, the Marsden Railway landscape was seeing further changes with new houses springing up along the inland flank of the track all the way to the Marsden quarries, this housing being put up at the expense of the (by now rather dated and cramped) miners' cottages at Marsden village. Demolition of the village therefore began in the late 1950s and was so thorough and complete that to this day Marsden is known locally as 'the village that vanished'. In 1958,

A view from the Grotto footbridge with the mouth of the River Tyne and the North Sea seen off to the right. No. 506 (Hunslet 6617) is seen heading a train to Westoe on 11th April, 1968. By this date the conveyance of Marsden limestone was in the hands of road traffic with the resultant tipper lorries very much in evidence. *Ian S. Carr*

As part of the Marsden rundown, the former crusher sidings adjacent to Lighthouse bridge were used for storage of condemned stock. One item awaiting disposal was this ex-GNSR brake third six-wheeler. Although it is in dreadful condition, there is much of interest in this view of the guard's compartment end. The date is 16th April, 1968. *Ian S. Carr*

the trolleybus overhead equipment was dismantled along the coast road due to encroaching corrosion that plagued it during the quiet winter months.

As for Whitburn Colliery, this soldiered on into the 1960s. However, several events occurred during 1965 which raised question marks over the future of the pit.

The first shockwave came in May when it was announced that all major repairs of locomotives would cease at the Area central workshops. Then, in September, all rail traffic ceased at the two Marsden quarries still in production, the NCB selling the pair to the private quarrying company Slater Brothers & Co. in the same month. The little fleet of narrow gauge engines had worked right up to the day of the takeover only to be scrapped or sold off by the new concern.

The year 1965 also saw the arrival of diesels with three brand new standard gauge Hunslet designs arriving on the Marsden Railway between August and November. This trio allowed the NCB to downgrade the status of the remaining five steam engines thus:

Locomotive	*Status*	*Fate*
No. 2 (Hunslet 3191)	Spare	Sent to Lambton December 1965
No. 3 (Robert Stephenson, Hawthorn 7339)*	Spare	Sent to Washington 'F' pit March 1966
No. 5 (Robert Stephenson, Hawthorn 7132)	Spare	Sent to Boldon Pit December 1965
No. 9 (Robert Stephenson, Hawthorn 7749)	Stored	Scrapped April 1968
No. 10 (Robert Stephenson, Hawthorn 7811)	Stored	Scrapped April 1968

As can be seen the spare locomotives were quickly dispatched to other destinations following the dieselisation of the Marsden Railway, leaving just No. 9 and No. 10 (a sorry pair) dismantled and forgotten behind Whitburn shed.

The quick dispatch of the remaining spare steam engines proved to be a somewhat rash decision as the summer of 1966 saw a seamen's strike with the Harton Staiths shut down for the duration. This meant 24 hour days and 7 day weeks for the diesels with No. 509 in particular putting in sterling work as the train engine. Delivered on 13th May, 1966, this Barclay locomotive had already clocked up 1,547 miles by 4th July and had not been inside Whitburn shed for over a month. With the staiths shut, Whitburn pit was chosen as a stockpiling railhead for all the coal produced by the entire Harton group, hence the intensive activity. However, this vast increase in traffic proved to be an Indian summer for the Marsden Railway, and bleak days lay ahead.

In October 1967 a rumour spread like wildfire amongst the Whitburn colliers that both Harton and Whitburn pits were earmarked for closure as part of the general rundown of the coal industry. Then in January 1968 NCB management warned the workforce that a weekly target of 6,500 tons of coal had to be reached to avoid closure. By 21st May the miners managed to exceed the target by 41 tons. Two weeks later the pit was shut down and 819 men were put out of work.

The last shift came to the surface on Friday 7th June, 1968 with both the pit and the Marsden Railway officially abandoned on 8th June pending the removal of the last train loads of coal up the branch to Westoe. The news spread fast and two enthusiasts' bodies of the day, the Manchester Locomotive Society, and Stephenson Locomotive Society, hastily arranged a railtour to take in not just the Marsden Railway but the entire Harton system. The 'Durham Rail Tour' thus arrived in rather grand style behind the *Flying Scotsman* at Tyne Dock station, with the 250 enthusiasts

* Although this engine was allocated the number '3', the number was never actually carried on the locomotive.

On 17th April, 1968, No. 505 (Hunslet 6616 of 1965) crosses Redwell Lane bridge on its way down to Whitburn Colliery. The original bridge was replaced in 1937 by this concrete version while the Austin 'A40' passing beneath, clearly dates this picture. *Ian S. Carr*

No. 505 makes its final approach toward the redundant Whitburn Colliery station with the 'Durham Coast Railtour' on 7th September, 1968. Despite all the railway infrastructure still being in place, as far as the NCB was concerned, both pit and branch line had ceased to exist. *Ian S. Carr*

This 24th October, 1968 photograph captures what must surely be one of the very last trains to run on the Marsden Railway. The road crane has already lifted the track from the station area (*seen in the distance*) with the 'face' on the colliery wall looking suitably horrified! *Ian S. Carr*

then transferred into a train of open wagons at Dean Road sidings headed by an English Electric locomotive for a tour of the electrified system. The NCB electric was exchanged with a pair of NCB diesels at Westoe (in 'top and tail' formation) for the journey down to the deserted Whitburn Colliery. Thus No. 505 (the leading Hunslet) became the last locomotive officially to run down the coast to Marsden, although subsequent locomotives were used of course in later track clearing operations.

The demolition of Whitburn Colliery began in October 1968. Aptly, the lifting of the Marsden Railway began at the same time and the rails were removed all the way up the coast as far as the abandoned Mowbray Road signal box where the remaining track was still employed as part of the thriving electrified system. This track lifting left two surviving railway facilities at Whitburn high and dry, the Landsale coal depot and the Area central workshops, and these said facilities were now served by road. As for the remaining 86 acres of land which once comprised Whitburn Colliery, this was now wasteground.

The following years saw the sweeping away of the final remnants of what was once the Marsden Railway. The Lighthouse bridge was removed in 1969 and Redwell Lane bridge in 1974. All the signalling equipment including the boxes of course soon disappeared as did the little platform at Marsden Cottage. By the end of the 1970s, only Westoe Lane still stood. Continuing in its role as offices, it remained curiously detached from the electrified remains of the Marsden Railway until it too was demolished *circa* 1980. The annexed, and by now totally remote, Area central workshops continued to function in a limited capacity until it too was closed down in 1985, for the past 17 years it had survived without a rail link.

A fascinating little display of 1960s plant vehicles make slow but steady work of removing Lighthouse bridge in this 22nd January, 1969 photograph looking back up the coast road toward South Shields. The signal set at danger is almost humourous. *Ian S. Carr*

The 11th April, 1988 saw NCB track gangs upgrading the remaining headshunt of what was once the SSMWCR in readiness for the introduction of British Rail-hauled trains into Westoe. The AEG locomotive No. 9 awaits removal to the Tanfield Railway for a well earned rest after putting in 75 years service on the Harton Railway. Note Westoe's lofty Crown Winder looming over the rooftops to the right. *Ian S. Carr*

Only the original ½ mile section of the Marsden Railway running from Whitburn Junction and Dean Road sidings, up Chichester Road Bank, past Westoe Lane, and on to the headshunt just south of Westoe Colliery continued to function as part of the electrified system. Although bereft now of its little passenger trains, steam locomotives, and station, Westoe Lane continued to be a scene of some fascination in the 1980s with German electric locomotives of Edwardian vintage still operating alongside later English designs. In 1989 however, the electrification of the former Harton system was abandoned as Westoe Colliery began to be served directly by British Rail, which ran its own 36-wagon trains from the pit, down Chichester Road Bank and out through Whitburn Junction to the new Tyne Coal terminal situated at Tyne Dock. At 3,250 horse power and 125 tons, the BR class '56' diesels which hauled these lengthy trains were by far the heaviest and most powerful locomotives to run on Marsden metals and made an impressive sight as they came to grips with the 1 in 39 section of Chichester Road Bank. By 1990 all other trackwork on the Harton system had been abandoned and so this remaining spur, the oldest section of the entire railway, dating as it did back to 1879, had outlived what was once a formidable and extensive system.

In this final format, the Marsden Railway was to last just four years and, on 7th May, 1993 Westoe Colliery, the last pit on the River Tyne, was to shut down forever. In terms of the wholesale decimation of the coal industry which was taking place at this time, Westoe was seen as just one more statistic. For the Westoe workforce (the last of the Tyneside miners) it was devastating. With no private buyers on the horizon, Westoe Colliery was demolished in April 1994 and the final section of the Marsden Railway lifted.

The remaining stump of the former SSMWCR the section which served as a headshunt was still in active use in April 1988 even though the overhead wires had been removed by this time. Barclay 659 turns the corner at Mowbray Road but will only travel for a few more yards before reversing its NCB train into Westoe sidings which are evident on the right. *John Furness*

Chapter Eight

Remains

The trackbed of the Marsden Railway offers the industrial archeologist some reward today. A round trip of 7 miles is something of a challenge, but two rather intriguing public houses *en route* provide refreshment and the whole system can of course be accessed by car courtesy of the coast road.

The site of Westoe Lane station (grid reference NZ370666) is now, alas, obscured by housing. However a grass bank topped by a limestone wall sandwiched between the houses and Chichester Road marks the southern flank of the railway where it ran through the station, the so-called 'Wall side'. To the west beyond the houses, Chichester Road drops away back to South Shields and provides a good indication as to the steepness of the former Chichester Road Bank. To the east of the station site at the end of Iolanthe Terrace, two pillars straddling a footpath mark the entrance of what was once the original footbridge, while a bricked section of walling immediately adjacent to this indicates where the Westoe Colliery entrance once stood. Beyond the wall the site of Westoe Colliery is now an area of desolate high ground. Here, the crown shaft has been capped in concrete and provided with an air-vent. The Marsden Railway trackbed curves away to the south here hugging the edge of the houses. This area is inaccessible to the visitor, but a walk down Mowbray Road reveals where the railway once crossed the road. By continuing down Mowbray Road towards the sea, the visitor will come across 'the Marsden Rattler' public house complete with a clock tower which came originally from Manors station in Newcastle, and two coaches which are disappointingly not of 'Rattler' origin but ex-British Railways and LMS stock.

The trackbed continues southward from Mowbray Road and runs parallel to the back of the former miners' houses in Tadema Road and this indicates how far the Marsden Railway ran after the bulk of it had been lifted in 1968. Close to the end of this headshunt, some brick foundations can just be discerned amongst the weeds, the remains of Mowbray Road signal box.

Post-1968 housing now obscures the continuing track bed but an alleyway can be followed through the housing. Trow Rocks footbridge was once situated where this alleyway crosses another footpath at right angles. The alleyway continues along the course of the railway until the coast road is reached.

Here it is much easier to define the original course of the railway and the track bed is indicated by a bank of grass sandwiched between the houses and the coast road. There are pleasant views here across 'the leas' towards the sea.

Continuing along the trackbed, a bus stop is reached where the coast road is seen to curve almost imperceptibly inland beneath an area of high ground and this is where Marsden Cottage Halt once stood. The trackbed then crosses Redwell Lane where the remains of the bridge abutments can still be seen. Beyond this, the coastal housing finally runs out and a caravan site is reached. Between the caravan site and the golf course, the ground is seen to climb away inland indicating the route of the spur which ran up to the Marsden Old

Quarry. The quarry itself survives to this day, overgrown but not infilled. On the opposite side of the coast road towards the sea, the entrance to the Marsden Grotto is reached. The Grotto itself is hidden from view at the foot of the cliffs but is well worth a visit as it represents one of Britain's most unusual public houses.

Continuing along the railway trackbed, the Marsden lime kilns are reached. Although fenced off, the kilns are surprisingly intact and are in fact preserved by National Heritage. They are both impressive and immense and the later brick-built versions can be seen at the southern end of the original limestone battery. The iron belts which straddle the kilns held the brickwork together in the extreme heat when the kilns were being fired. They were finally put out in the 1950s. Behind the kilns, the two former Marsden Lighthouse quarries are still worked as one extensive single operation and are out of bounds to the visitor. The trackbed continues here between the coast road and the quarry but it is difficult to discern (although the original Marsden village Post Office still stands and is now a private dwelling). Souter Point lighthouse dominates the scene here and, as the first lighthouse in the country successfully to use electric light, it is preserved by the National Trust, as is the adjacent foghorn - for so long the bane of visiting football teams!

Here some railings flank the seaward side of the coast road, and this indicates where Marsden village school once stood. The school has long gone, along with the rest of Marsden village (the site of which is now covered by grassland all the way to the cliff edge). On the inland side of the coast road the abutment of the former Lighthouse bridge remains largely intact while the adjacent stonework was in fact a retaining wall to support the crusher sidings which at one time lay directly above. On the other side of the road the trackbed now disappears where it once entered Whitburn Colliery and the whole pit area is a scene of landscaped ground complete with Nature Reserve, although the heart of the former complex is now marked by a beacon. Approaching this, the visitor comes to two heavy pillars adjacent to the coast road which now serve as the entrance to a public house car park. These at one time guarded entry into Whitburn Colliery itself.

The streets of miners' houses which make up Whitburn Colliery village still stand to this day on the inland flank of the coast road. Now bereft of the pit it once served, the village appears bleak indeed. However, on quiet evenings when it is possible to hear the sea hissing over the rocks beneath the cliffs, the sounds and the darkness combine to conjure up former days when the little train ran along the cliff-top to the pit by the sea.

Appendix One

The Marsden Railway Locomotive Fleet List

No.	*Builder*	*Works No.*	*Built*	*Type*	*On SSMWCR*
1	Manning, Wardle			0-6-0ST	1879-1895
2	Black, Hawthorn	504	1879	0-6-0ST	1879-1905
3	Black, Hawthorn	716	1882	0-6-0ST	1882-1910
4	Black, Hawthorn	826	1884	0-6-0ST	1884-1948
5	Robert Stephenson	2629	1887	0-6-0ST	1887-1922
7	Chapman & Furneaux	1158	1898	0-6-2T	1898-1912
8	Sharp, Stewart	1501	1864	2-2-2WT	1899-1907
9	Blyth & Tyne Railway		1862	0-6-0	1900-1913
10	Blyth & Tyne Railway		1862	0-6-0	1900-1914
6	Robert Stephenson	2160	1874	0-6-0	1908-1912
8	Robert Stephenson	1973	1870	0-6-0	1908-1929
11	Manning, Wardle			0-4-0ST	1908-1920
6	Robert Stephenson	2056	1872	0-6-0	1912-*c.*1927
10	R. & W. Hawthorn	1564	1873	0-6-0	1914-1931
6	North Eastern Railway		1882	0-6-0	1927-1930
5	North Eastern Railway		1881	0-6-0	1929-1953
6	Robert Stephenson	2587	1884	0-6-0	1930-1935
7	North Eastern Railway	38 [1892]	1892	0-6-0	1930-1935
8	North Eastern Railway	3 [1889]	1889	0-6-0	1931-1954
1	Andrew Barclay	1639	1923	0-6-0ST	1931-1960
7	North Eastern Railway	43 [1889]	1889	0-6-0	1935-*c.*1938
6	North Eastern Railway	23 [1889]	1889	0-6-0	1935-1951
7	North Eastern Railway	631	1898	0-6-0	1939-1946
9	Hunslet	3191	1944	0-6-0ST	1947-1965*
10	Robert Stephenson & Hawthorn	7339	1947	0-6-0ST	1947-1966
1513	Hudswell, Clarke	1513	1924	0-6-0ST	1948 & 1950
5	Robert Stephenson & Hawthorn	7132	1944	0-6-0ST	1948-1965*
6	Robert Stephenson & Hawthorn	7603	1949	0-6-0ST	1949-1964*
4	North Eastern Railway		1883	0-6-0	1950-1952
7	Robert Stephenson & Hawthorn	7695	1951	0-6-0ST	1951-1965*
9	Robert Stephenson & Hawthorn	7749	1952	0-6-0ST	1952-1968*
10	Robert Stephenson & Hawthorn	7811	1954	0-6-0ST	1954-1968*
4	North Eastern Railway	28 [1897]	1897	0-6-0	1956
4	Robert Stephenson & Hawthorn	7294	1945	0-6-0ST	1959-1963*
505	Hunslet	6616	1965	0-6-0DH	1965-1968
506	Hunslet	6617	1965	0-6-0DH	1965-1968
507	Hunslet	6618	1965	0-6-0DH	1965-1966
509	Andrew Barclay	514	1966	0-6-0DH	1966-1968

* Intermittent periods also spent at Boldon Colliery (see the main text).

Appendix Two

A Summary of Production at Whitburn Colliery 1881-1961

Date	Tons drawn	Averages	Days worked
1881	9,222	-	-
1882	469,458	-	-
1883	132,912	486	273
1884	145,466	533	270
1885	149,496	593	252
1886	178,024	792	225
1887	208,673	784	266
1888	199,082	774	257
1889	242,708	890	272
1890	292,631	1,102	265
1891	302,283	1,179	256
1892	229,435	1,172	195
1893	337,642	1,459	231
1894	434,160	1,789	242
1895	450,464	2,060	218
1896	521,379	2,294	227
1897	633,537	2,470	256
1898	686,091	2,599	263
1899	629,502	2,396	261
1900	647,233	2,461	262
1901	591,540	2,281	259
1902	416,935	1,522	273
1903	480,952	1,743	275
1904	505,236	1,806	279
1905	504,026	1,832	275
1906	490,664	1,761	278
1907	502,461	1,849	271
1908	551,407	1,977	278
1909	591,344	2,134	277
1910	626,226	2,372	264
1911	608,145	2,318	262
1912	547,434	2,299	238
1913	575,899	2,195	262
1914	534,901	2,186	244
1915	501,709	1,894	264
1916	509,310	1,904	267
1917	442,383	1,915	230
1918	403,829	1,642	245
1919	420,934	1,668	252
1920	444,451	1,777	250
1921	366,957	2,044	179
1922	566,262	2,337	242
1923	565,104	2,350	240
1924	571,632	2,317	246
1925	319,014	2,274	140
1926	237,264	2,288	103
1927	606,010	2,396	252
1928	528,006	2,084	253
1929	655,868	2,459	266
1930	652,448	2,592	251
1931	600,288	2,357	254
1932	479,341	1,864	257
1933	397,384	1,667	238
1934	528,219	2,087	253
1935	548,953	2,064	265
1936	452,262	1,683	268
1937	503,613	1,905	267
1938	511,669	1,967	260
1939	478,654	1,836	260
1940	449,958	1,772	253
1941	356,670	1,371	260
1942	324,682	1,238	262
1943	386,317	1,485	260
1944	375,973	1,483	253
1945	361,434	1,383	261
1946	452,797	1,708	265
1947	484,043	1,877	258
1948	520,809	2,019	258
1949	520,375	2,191	237
1950	540,289	2,170	249
1951	593,858	2,208	269
1952	547,980	2,092	262
1953	537,546	2,076	259
1954	518,872	2,114	245
1955	537,131	2,031	264
1956	590,859	2,290	258
1957	521,747	-	-
1958	455,103	-	-
1959	481,440	-	-
1960	425,509	-	-
1961	498,998	-	-

Coals drawn and averages per day are to the nearest ton, days worked are to the nearest whole day; averages and days worked were only recorded between 1883 and 1956, tons drawn were only recorded up to 1961.

Coal was drawn for 11 hours per day until 1890, 10 hours per day between 1890 and 1919, with the 7 hours Act taking effect from 14th July of that year. Miners worked 10 days per fortnight from 29th January, 1921.

National strikes occurred between 1st March-10th April, 1912; 18th October-4th November, 1920; 1st April-9th July, 1921; and 1st May-6th December, 1926.

Whitburn was also shut down due to a local dispute 23rd April-28th September, 1925.

Appendix Three

A list of final tickets sold at Westoe Lane station, November 1953

Ticket No.	*Type*
1293	First class adult return to Marsden at 9*d*.
23	First class child return to Marsden at 4½*d*.
9558	Third class adult return to Marsden at 6*d*.
8734	Third class child return to Marsden at 3*d*.
5077	Workmen's weekly pass at 3*d*.
5241	School children's ticket at 1½*d*.
2113	First class adult single to Marsden at 6*d*.
27	First class child single to Marsden at 3*d*.
9376	Third class adult single to Marsden at 4*d*.
858	Third class child single to Marsden at 2*d*.
2546	Adult single from Marsden Cottage Halt at 2*d*.
210	Single ticket for dog to Marsden at 3*d*.
250	Single pram/bicycle ticket to Marsden at 2*d*.

Note:
These were the final tickets sold while the passenger service was in operation. Further tickets continued to be sold from Westoe Lane station for the benefit of collectors after November 1953.

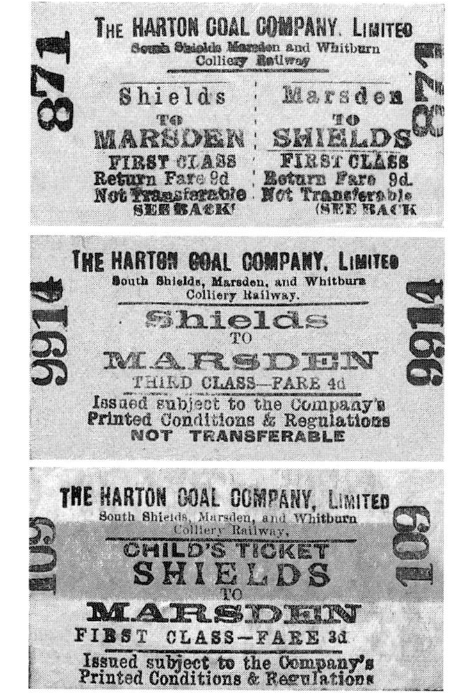

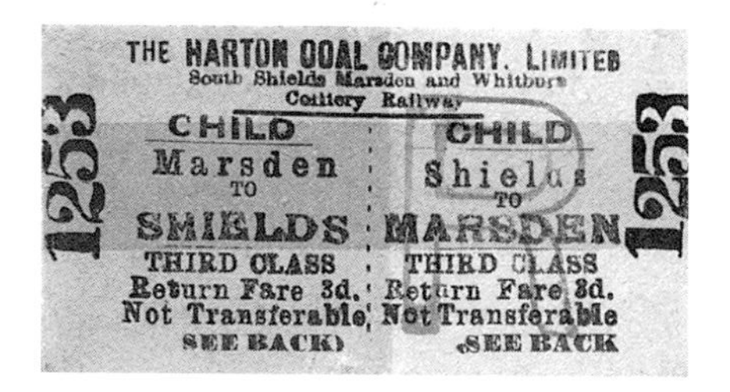

Westoe Lane ticket receipts taken in the final years of operations

Total receipts for 1949
£5 8s. 0*d*. (returns), £3 11*s*. 3*d*. (singles).

Total receipts for 1950
£5 2*s*. 9*d*. (returns), £4 5*s*. 6*d*. (singles).

Total receipts for 1951
£3 9s. 1½*d*. (returns), £3 6*s*. 0*d*. (singles).

Total receipts for 1952
£4 18s. 9*d*. (returns), £3 17*s*. 1*d*. (singles).

Total receipts for 1953
£7 16s. 10½*d*. (returns), £6 6*s*. 8*d*. (singles).

Acknowledgements

J.P.R. Bennett, M.N. Bland, Fred Bond, Geoff Burrows, Ian S. Carr, R.M. Casserley, John Clayson , J.B. Dawson, John Dixon, Keith Fenwick, Nick Fleetwood, John Furness, Alison Hatcher, Billy Henderson, Industrial Locomotive Society, Industrial Railway Society, Les Jackson, Frank Jones, R.W. Kidner, A.P. Lambert, Sydney A. Leleux, Debbie Lewis, P.J. Lynch, Sandy Maclean, Christine Matton, David Monk-Steel, C.E. Mountford, Jim Peden, Ken Plant, Peter Rowbotham, Ian Scrimgeour, Mrs F. Shepherd, Signalling Record Society, Neil Sinclair, William J. Skillern, Mark Smith, Martin Smith, Les Snaith, Alan Snowden, South Shields Central Library, Neville Stead, South Tyneside Metropolitan Borough Council Development Services Dept, John Talbot, Kenneth L. Taylor, Dr Bob Tebb, Jack Thompson, John Watling, Russell Wear, Helen Wiltshire. And lastly special thanks to Robert F. Wray for his constant correspondence and encouragement.

Index